Walzer, Just War and Iraq

In recent years, questions of ethical responsibility and justice in war have become increasingly significant in International Relations (IR). This focus has been pre-cipitated by the invasions of Afghanistan and Iraq led by the United States (US). In turn, Western conceptions of ethical responsibility have been largely informed by human rights-based understandings of morality.

This book directly addresses the question of what it means to act ethically in times of war by drawing upon first-hand accounts of US fighting in Iraq during the 2003 invasion and occupation. The book focuses upon the prominent rights-based justification of war by Michael Walzer. Through an in-depth critical reading of Walzer's work, this title demonstrates the broader problems implicit to human rights-based justifications of war and elucidates an alternative account of ethical responsibility: ethics as response.

By putting forward a compelling case for people to remain troubled and engaged with questions of ethical responsibility in war, this work will be of great interest to students and scholars in a range of areas including international relations theory, ethics and security studies.

Ronan O'Callaghan is a lecturer in International Relations at the University of Central Lancashire. His research interests include ethics and war, poststruc-tural philosophy, the politics of violence, and contemporary forms of political resistance.

Interventions

Edited by Jenny Edkins, Aberystwyth University and Nick Vaughan-Williams, University of Warwick

The series provides a forum for innovative and interdisciplinary work that engages with alternative critical, post-structural, feminist, postcolonial, psychoanalytic and cultural approaches to international relations and global politics. In our first 5 years we have published 60 volumes.

We aim to advance understanding of the key areas in which scholars working within broad critical post-structural traditions have chosen to make their interventions, and to present innovative analyses of important topics. Titles in the series engage with critical thinkers in philosophy, sociology, politics and other disciplines and provide situated historical, empirical and textual studies in international politics.

We are very happy to discuss your ideas at any stage of the project; just contact us for advice or proposal guidelines. Proposals should be submitted directly to the series editors:

- *Jenny Edkins (jennyedkins@hotmail.com) and*
- *Nick Vaughan-Williams (N.Vaughan-Williams@Warwick.ac.uk).*

As Michel Foucault has famously stated, 'knowledge is not made for understanding; it is made for cutting'. In this spirit The Edkins–Vaughan-Williams Interventions series solicits cutting edge, critical works that challenge mainstream understandings in international relations. It is the best place to contribute post-disciplinary works that think rather than merely recognize and affirm the world recycled in IR's traditional geopolitical imaginary.

Michael J. Shapiro,
University of Hawai'i at Manoa, USA

Critical Theorists and International Relations
Edited by Jenny Edkins and Nick Vaughan-Williams

Ethics as Foreign Policy
Britain, the EU and the other
Dan Bulley

Walzer, Just War and Iraq

Ethics as response

Ronan O'Callaghan

Routledge
Taylor & Francis Group

LONDON AND NEW YORK

First published 2016
by Routledge
2 Park Square, Milton Park, Abingdon, Oxon OX14 4RN

and by Routledge
711 Third Avenue, New York, NY 10017

Routledge is an imprint of the Taylor & Francis Group, an informa business

First issued in paperback 2021

British Library Cataloguing in Publication Data
A catalogue record for this book is available from the British Library

Library of Congress Cataloging in Publication Data
O'Callaghan, Ronan.
Walzer, just war and Iraq / Ronan O'Callaghan.
pages cm. — (Interventions)
1. Iraq War, 2003–2011—Moral and ethical aspects. 2. War—Moral and
ethical aspects. 3. Just war doctrine. 4. Walzer, Michael. I. Title.
DS79.767.M67O43 2015
172'.42—dc23
2015018918

ISBN: 978-1-138-93388-0 (hbk)
ISBN: 978-1-03-209824-1 (pbk)
ISBN: 978-1-315-67552-7 (ebk)

Typeset in Times New Roman
by Swales & Willis Ltd, Exeter, Devon, UK

Contents

Abbreviations

COIN	Counterinsurgency
CPA	Coalition Provisional Authority
IED	Improvised Explosive Device
IGC	Iraqi Governing Council
IR	International Relations
JWT	Just War Tradition
PTSD	Post-traumatic Stress Disorder
SCIRI	Supreme Council for the Islamic Revolution in Iraqi
TAL	Transitional Administrative Law
UIA	United Iraqi Alliance
UN	United Nations
US	United States of America
WMDs	Weapons of Mass Destruction

Acknowledgements

To say that bringing this book to term was an unambiguously enjoyable experience would be somewhat insincere. When any project dominates your life for a sizeable length of time, let alone the best part of six years, you have to accept the agathokakological nature of the beast. The primary reason that this venture has proved worthwhile and, ultimately, successful is due to the help and understanding of a number of fantastic people.

The vast majority of research for this book was conducted at the University of Manchester during the development of my PhD thesis, where I was lucky to be surrounded by a group of intelligent and supportive colleagues. In this respect, my supervisor Professor Maja Zehfuss deserves irreplicable plaudits for her guidance and patience. Maja has aided this project since its inception, ensuring that what you have before you today is honest, engaging and, above all, moderately legible. Her incremental assistance has shaped this book, and indeed this author, profoundly. In turn, I would like to sincerely thank my second supervisor, Dr Peter Lawler. Peter's keen eye for detail and wide breadth of knowledge of International Relations literature helped to ground this work, and it is undoubtedly better for his input.

In addition, I would like to thank Dr Nick Vaughan-Williams for his faith in me, and support during the last year. Nick's help has been invaluable, and without it this book would not have made it to press. I would also like to thank Dr Véronique Pin-Fat for her support. On top of housing one of the most unique critical minds, Véronique remains one of the most charming and unpretentious human beings I have had the pleasure of meeting. My fellow PhD students Gareth Price-Thomas, Jamie Johnson, Julia Welland, Róisín Read, Henrique Furtado, Thomas Gregory, Tom Houseman, Oliver Turner, Kathryn Starnes, Astrid Nordin, Rebecca Ehata, Sara Marie-Siobhán Kallock, Rachel Massey, Emmy Eklund, Tomas Maltby, Aggie Hirst, Giulia Sirigu, Rob Munro, James Alexander, Guro Buchanan, David Tobin and Dean Redfearn made writing this book a far more manageable and rewarding experience. I feel lucky to count you all as friends. I would also like to thank my current colleagues at the University of Central Lancashire, in particular Dr Antonio Cerella, for helping me push this book over the final hurdles. Thanks must also go to Dr Cian O'Driscoll for his interest and support of my work during the last few years. His help, advice and friendship has been greatly appreciated.

Above all, it is my partner, Laura White, who must be acknowledged. Her love and support throughout the last six years has been the single most valuable contribution to this project. Without her this book would not have been possible. Gratitude must also go to my family, my parents Tomás and Mary O'Callaghan, and my siblings Cian and Orla O'Callaghan. They have loved and supported me throughout my life, and I feel privileged to have grown up around such wonderful people. Finally special thanks must go to my grandmother Lily O'Callaghan and my recently departed grandfather Paddy-Joe O'Callaghan. You will struggle to find two more loving, kind and compassionate people.

Introduction

The Tower of Babel and the ideal of unity

The story of the Tower of Babel is recounted in the book of Genesis (Genesis 11: 1–9). The tale attempts to provide a mythological account of the emergence of different languages and, thereby, different cultural groupings. The narrative depicts an existence in which all the world's people spoke the same language and settled in the town of Shinar. The people of Shinar soon set about building a city, the centrepiece of which was a giant tower that reached toward the heavens. God, however, visited the city and decided that a people so thoroughly unified would have nothing left to reach for: they would become Gods in their own eyes and would therefore have no need for him. In response, God confounded the people of Shinar's speech so that they could no longer understand each other. Confronted by their lack of mutual understanding they ceased building the city, the tower fell to ruin and its populace was scattered across the world. This story is one of many mythological accounts that link social unity with the ideal of common language. In the tale of Shinar we start with unity that is violently split and divided into difference. While the story aims to account for difference, it nevertheless attempts to do so by describing it in terms of a fall or corruption: God feared that the people of Shinar were reaching for the heavens, his rightful place, and punished them by assailing the foundation of their community – their common language.

The tale of Shinar, or Babylon as it is now more commonly known, in some respects displays a sense of perverse symmetry with the modern history of the nation that houses its ruins: Iraq. Since the 1960s Iraq's diverse populace was unified, not through language, but through the militaristic rule of the Ba'ath regime. Ba'athism, in other words, provided an antithesis to the free association of the citizens of Shinar, a repressive form of forced unity. In 1979 when Saddam Hussein assumed autocratic rule of the nation, this semblance of unity became further entrenched through dogmatic laws and the repression of any Iraqi who opposed Ba'athist ideals. In the decades that followed Saddam's ascension, Iraq's oil wealth dwindled and modern infrastructure crumbled in the face of a series of wars – with Iran, an international alliance during the Gulf War, and internal conflicts against disenfranchised Shi'a and Kurdish factions – and a prolonged regime of international sanctions. While the nation fell into economic disrepair and its populace was ravaged by the effects, its unifying centre of military rule remained largely unscathed. The 2003 invasion of Iraq led by the United States of America (US)[1], however, removed the unifying centre of Ba'athist repression and scattered the Iraqi people increasingly into entrenched groupings based upon their ethno-sectarian identity. The Iraq we know today, over a decade later, is defined by its differences: the largely autonomous Kurdish population in the north seeking to insulate itself from the rest of the country; the Shi'a majority who control Iraq's governing institutions and political programme; and the disenfranchised Sunni minority who have increasingly turned toward violence as a means to exert their influence. And yet, the dissolution of Iraq's unity does not resemble Shinar's mythical fall from grace. It is not the case that Saddam's Iraq was a loving unity torn apart by fear and confusion into alterity by foreign intervention. Instead, Iraq represents the greatest fear of contemporary humanitarian thinking: the realisation that intervening in the defence of human rights does not necessarily solve the problem and can, perhaps, exacerbate the conditions for people on the ground. The modern history of Iraq thereby illustrates that the great humanitarian benevolence of liberating oppressed people can, conversely, serve as the most severe of punishments. Nonetheless, humanitarians could soothe themselves in the self-affirmed knowledge that Iraq was not a proper intervention: it was an illegal US war driven by strategic and economic self-interest, not a genuine defence of human security.[2] As such, Iraq didn't trouble the humanitarian coupling of rights and intervention. What I want to suggest, in contrast, is that Iraq cuts at the very core of the tensions facing current humanitarian ideals and conceptions of ethical responsibility.

Some wars seem to occupy important positions in the cultural memories of particular groups of people. As a Westerner born in the 1980s, Iraq has become the defining symbol of war in my lifetime. Although I have vague memories of the Gulf War in 1990, the 2003 invasion coincided with the start of my university education and, as such, represented the first time that I began to think critically about the ethical and political implications of war. To most of my peer group and I (undoubtedly influenced by Noam Chomsky), the US invasion of Iraq was an amoral and illegal attack upon Iraq and its people. Many of us took to the streets, engaged in demonstrations and tried desperately to stop this unjust war. However, there was a simultaneous recognition that Saddam Hussein's rule was predicated upon the brutalisation of a large proportion of the Iraqi people. The fact that the US was morally wrong in its actions did not nullify the injustices inflicted by the Ba'ath regime; it did not undo the mass repression of Iraqis, the systematic state torture, the mass starvation witnessed under the sanctions regime, or justify the thousands of Kurds and Shi'a killed throughout the 1980s and 1990s. The 2003 invasion, in this respect, retained a pervasive ethical tension: my rejection of the US justifications for the war was coupled with the knowledge that ordinary Iraqis were being repressed by their government on a daily basis. This was a morally wrong war against a morally wrong regime, and Iraqi lives and liberties would be violated regardless of outcome. For this reason, Iraq posed difficult ethical decisions. Did we have a responsibility to save Iraqis from Ba'ath repression? Did we have a responsibility to stop the US war? And how could we condemn the US war without endorsing the Ba'ath government and its persistent violations of human rights? Iraq, in this way, brought together the problematic relationship between humanitarianism and the spectre of war. Perhaps more importantly, Iraq illustrated how politicians could justify, or at least attempt to justify, war as a defence of human rights. For how could we possibly defend the rights of Iraqi people without violent intervention?

There has already been a rich critical response to the Iraq War from scholars of International Relations (IR). These responses have been crucial to furthering our understanding of the conflict and helping us to unpack the problematic dimensions of modern war and its justifications. Chomsky (2004), for example, argues that Iraq represents the symbolic reassertion of US power on the global stage; John Mearsheimer and Stephen Walt (2008) contend that Iraq provides evidence of the US's misguided foreign policy programme in the Middle East, driven by its alliance with Israel; and David Harvey (2005) points toward the economic dimensions that influenced the 2003 invasion. Perhaps most interestingly, Derek Gregory (2004) presents the invasion of Iraq as a component of a much larger contemporary colonial project. Gregory's central argument is that the invasion is driven by logics and strategic goals that tie into the broader aim of enforcing Western moral, political and legal norms throughout the Middle East. While critical responses have undoubtedly expanded our knowledge and understandings of US motivations and the wider socio-political contexts surrounding the invasion, they do not necessarily help us resolve some of the most pressing ethical questions that Iraq raises. Fundamentally, they have very little to say about the tension between the ethical desire to help oppressed people and the implications of using war as a tool to achieve this aim. In short, they explain the problems of the US war but tell us very little about how this can help us respond to future humanitarian crises.

This book is partly a response to my own anxieties about the ways in which our sense of ethical responsibility toward other people can help to justify particular forms of violence. In short, I am interested in how humanitarian ideals have become a key component in the justification of contemporary violence. This book is primarily concerned with the relationship between the moral justifications of war and the consequences that follow from these justifications. In this respect, it is an attempt to open discussions on justice and war to alternative understandings of what it means to act ethically. This book makes three main contributions to contemporary IR. First, it provides a robust critique of human rights-based conceptions of ethical responsibility and justifications of war. Second, it elucidates an alternative understanding of ethical responsibility that follows from the work of Jacques Derrida. And third, it illustrates how this alternative account of ethical responsibility can help us understand and respond to the 2003 invasion of Iraq and its legacy. The overarching aim is to unpack a conception of ethical responsibility that challenges conventional models of wartime ethics that attempt to provide clear general moral rules. In response to these models, this book argues that we must reject the desire to seek ethical closure. Instead, we must move toward more sustained and active engagements with the consequences of violent actions without seeking definitive or conclusive ethical satisfaction.

Michael Walzer and the just war tradition

The purpose of this book is to critically reflect upon and challenge the relationship between humanitarian ideals and the justification of violence. In engaging with humanitarian justifications of war this book focuses upon the just war tradition (JWT) and, more specifically, on the work of Michael Walzer. The JWT provides a particularly useful way of thinking about the contemporary ethical dimensions of war, not least because it often posits itself as a middle ground between pacifism and realism: a conception of ethical responsibility that endorses the use of violence, in certain cases, as an instrument of justice. The JWT, in this sense, attempts to account for ethics and morality in instances – like Iraq – in which violent intervention appears to be the only effective means of addressing human rights violations. The JWT has a long and diverse history that is commonly claimed to originate with canonical figures like Marcus Tullius Cicero (Bellamy 2006) and St. Augustine of Hippo (Corey and Charles 2011). However, the vast majority of scholars within the JWT are keen to emphasise that the tradition represents a historically informed, ongoing debate about the ethical dimensions of war. James Turner Johnson, for instance, states that the JWT represents a historically cultivated body of moral wisdom rooted in Western ideals, institutions and experiences that provides an important synthesis between idealism and realism (2001: 23). The JWT, in other words, attempts to account for morality and justice in war through historical reflection and pragmatic analysis. Yet the tradition does not constitute a singular approach to the question of justice in war, nor does it represent a shared set of moral values and beliefs. Rather, as Cian O'Driscoll highlights, the tradition posits many different assumptions, conditions and commitments, and privileges different forms and means of moral knowledge (2008: 91–92). The JWT, as such, is fragmented and disagreements are replete within the tradition itself.

The contemporary terrain of the JWT is complex, difficult to define in any singular manner and infused with tensions. Contemporary debates within the tradition, in turn, relate to a broad range of issues: methodological approaches, conceptions of morality and the tradition's overarching purpose. Debates, for example, include: the role of religion, the use of history, the nature of morality, and so on. In short, the JWT regularly finds itself deeply embroiled in internal disputes that, as Johnson (2013) suggests, are often more vigorously contested than the external debates. Although this book strives to emphasise the diverse nature of the JWT, it is not concerned with unpacking the vibrant arguments that Johnson alludes to. In fact, it is precisely because the JWT often focuses on internal disagreements that this book attempts to engage just war protagonists with issues and arguments often absent from their debates. The present argument, in this respect, is not concerned with outlining current debates in the tradition and situating an alternative conception of ethical responsibility from within this framework. Instead, this book is concerned with presenting a sustained challenge to some of the most commonly shared assumptions within the JWT: the ideal of universalising moral rules; the principle of non-combatant immunity; the doctrine of double effect; and the underlying belief that we can achieve justice through war. This is not to say that the JWT has not broached some of these topics already (for example, see O'Driscoll, Williams and Lang (2013) and Brough, Lango and van der Linden (2012)). However, the difference between this book and responses from within the tradition is that the present discussion is focused upon questioning the discursive foundation of the JWT: the idea that justice is achievable through violent actions. Even in its less absolutist forms, just war thinking is designed to provide guidelines for making *relative* moral decisions (Johnson 2011: xxxiii). In other words, the JWT is founded on the pragmatic ideal of facilitating morally justifiable actions during war via the provision of rules, regulations, guidelines and norms. The fundamental difference between the arguments presented in this book and the JWT, therefore, is a rejection of the ideal of moral justification and ethical satisfaction. In contrast, the idea of *ethics as response* proposed in this work serves as a means to account for how people can act ethically in the absence of relative moral certainty.

Despite being considered a canonical figure in IR's understanding of just war theory, Michael Walzer and his work represent a distinctive approach within the broader JWT. Something must hence be said concerning why Walzer, and not another author or authors, was chosen for the basis of the present discussion on ethics and war. Walzer is a key figure in the re-emergence of the JWT within contemporary IR scholarship, particularly in the US context. According to Jean Bethke Elshtain, Walzer gave just war theory a new lease of life and put it back on the map of contemporary social and political theory (1992: 2). In turn, *Just and Unjust Wars*, Walzer's seminal work on the ethics of war, has been described as a 'modern classic' (Boyle 1997; Hendrickson 1997; Knootz 1997; Nardin 1997; and Smith 1997), and has become a standard – if not *the* standard – text on courses on morality and war (Pin-Fat 2010: 89–90). In addition to his importance in regard to academic discussion, Walzer's work has also been directly incorporated within US military doctrine and strategy guides. For example, in the 2007 US Army and Marine Corps Counterinsurgency (COIN) Field Manual, Sarah Sewall (a consultant on the manual) informs us that Walzer restored our ability to think clearly about war and explains that the manual aims to apply Walzer's conception of 'fighting well' to the terrain of counterinsurgency (Petraeus *et al.* 2007: xxii). Walzer's understanding of morally justified warfare, in this respect, has become a common component of academic and military discussions on war. Consequently, Walzer's conception of justice in war broadly fits within human rights discourses that, in varying ways, justify war in terms of the defence of individual or collective rights.[3] Walzer rejects the JWT's theological heritage, thereby distinguishing himself as a secular just war theorist. This does not imply that Walzer rejects the arguments presented by theologically orientated scholars. Rather, Walzer believes that the language of human rights provides a more appropriate and secularised moral vocabulary to discuss contemporary morality in war. Walzer, in this respect, approaches war from what he considers to be a legalistic paradigm; a position that Johnson contends is structurally similar to Hugo Grotius (2010: 79–80). In this sense, Walzer's understanding of war offers an exemplary – and robust – account of the relationship between human rights discourse and contemporary justifications of violence. Walzer attempts to provide moral justifications of war that are based on humanitarian ideals, secular in design and directly related to contemporary understandings of international law. While Walzer undoubtedly articulates a unique and particular account of this relationship, his exposition helps us to illustrate a number of more generalised problems in human rights-based justifications of war. Walzer, as such, is a seminal figure in the evolution of just war theory in the modern era: he offers an iteration of just war thinking that claims to be detached from its theological past and orientated specifically to help us address the ethical dilemmas of modern war.

There has been a number of influential critiques of Walzer's work in recent years. Jeff McMahan's *Killing in War* (2009), for instance, challenges Walzer's principle of 'moral equality between combatants' and the traditional just war separation of *jus ad bellum* (just recourse to war) and *jus in bello* (just conduct in war). More directly, Véronique Pin-Fat (2010) provides a critical examination of Walzer's depiction of ethics, subjectivity and politics from a broadly post-structural perspective. This book, however, moves beyond the recent critiques of Walzer's work. On the one hand, this project, as will be explained, rejects the rule-based system of morality that McMahan employs to justify war. On the other hand, while this book largely agrees with Pin-Fat's critique of Walzer, Pin-Fat does not engage with Walzer's just war theory or provide an illustrative account of how Walzer's morality relates to the 'real world'. In contrast, by situating the discussion on Walzer in the context of the Iraq War, this work begins to unpack the consequences of accepting some of the foundational just war assumptions and, in addition, takes a holistic approach to Walzer's conception of ethics. The majority of responses to Walzer treat his communitarian writings as distinct from his just war theory. By comparison, this book argues that Walzer's broader political arguments cannot be detached from his justifications of war: Walzer justifies war in defence of a political community's right to self-determination and, therefore, his justification is fundamentally intertwined with his conception of the political and community. This approach asks us to view Walzer's conception of justice, morality and ethics within the prism of his underlying political beliefs and motivations.

In the preface to *Just and Unjust Wars*, Walzer declares his intention to 'recapture the just war for moral and political theory' (2006a: xxii). What is paramount to Walzer's reclamation is the idea that critical judgements about war should not be the province of political leaders who often deploy violence as a means to achieve their own strategic ends. Instead, Walzer contends that ordinary people who suffer from war's consequences should be empowered to make moral judgements. In this sense, Walzer is arguing that questions of war should be democratic imperatives rather than the privilege of elite opinion.[4] Walzer's overarching ambition to stimulate active public discussions about war remains both admirable and desirable because it articulates the belief that people should be engaged and concerned with questions of ethics and war. The aim of this book is to demonstrate how the ways in which Walzer begins to think about war work counter to this ideal. Specifically, I will illustrate how Walzer's foundational assumptions – assumptions commonly shared by the wider JWT – militate against open discussions about ethics and war, and prevent us from identifying clear alternatives to just war's proposed middle ground solutions.

Deconstruction as methodology

This book employs a careful reading of Walzer's depiction of ethical responsibility as a means to posit an alternative understanding of ethical responsibility in war. This purpose follows Derrida's (1997) claim that deconstruction must strive toward productive readings that attempt to open discourses toward new approaches and understandings. In his writings, Derrida explains the tensions between the scholarly desire for accurate representation and the ramifications of interpretation. Derrida unpacks this through the concept of *non-transcendent reading*. Derrida argues that traditional philosophical criticism is concerned with uncovering the true meaning of a text: the critic reads a text carefully so that they can accurately represent and therefore critique it. Derrida suggests that the reason why critics focus on the ideal of a singular true meaning is due to fear that more interpretative readings would relax critical rigour and devalue their work (1997: 159). The serious scholar aims to faithfully represent a text in a rigorous manner, while interpretative readings allow scholars to frivolously say anything they like. As exemplified by so-called scientific methods of analysis, the perceived detachment and neutrality of the scholar is intimately linked to validity of their research: to admit that your work is sullied by interpretation is to risk discrediting it. Derrida calls this *transcendent reading*. He contends that this ideal assumes that reading allows for an unbroken transportation of meaning between author and reader: the meanings that the author intentionally embeds in a text can be fully uncovered by the reader through careful analysis (1988: 1). Derrida rejects this understanding of reading by drawing our attention to the necessary role played by interpretation. He argues that reading is an act in which both author and reader are implicated in a negotiated formulation of the text's meaning (Derrida 1992). Derrida explains that texts do not contain a singular truth or set of meanings waiting to be uncovered by diligent critics; rather, the reading of any text produces a singular, and modified, articulation of the text's intended meaning. In Maja Zehfuss's words, '[b]oth reader and writer are engaged in writing, together and simultaneously against each other' (2007: 25). The reconceptualisation of the act of reading positions the reader as an active participant in the construction of the text's meaning – what Derrida calls non-transcendent reading (1992: 44–45). Derrida, however, maintains that acknowledging the reader's role in the construction of a text is not equivalent to the relaxation of scholarly rigour. He explains that the reader must attempt to take account of the author's intended meaning while simultaneously rewriting the work through their own interpretation, thereby producing a singular and unique experience of the text (1992: 69–70). In other words, the reader's interpretation of the text both combines with and contests the author's intended meaning, producing a new account. Deconstructive methods, in this respect, aim to negotiate an opening between these dual responsibilities.

Deconstructions do this by demonstrating why the author is unable to accomplish what they want to accomplish, highlighting the fractures, tensions and inconsistencies contained within the internal logics of the text. In short, deconstructions illustrate why particular arguments undo themselves. In this book, for instance, I point toward an alternative understanding of the relationship between justice and violence by highlighting why Walzer's account of ethical responsibility fails to function in the way he wants it to.

Although Derrida contrasts non-transcendent reading with the ideal of transcendent readings, the former should not be understood as an alternative to finding the intended meaning. It is not a case that some readers locate the intended meaning and other readers produce new meanings. Instead, what Derrida is arguing is that every possible reading of a text is underpinned by deviation from the original intended meaning. In a short story on the concept of map making, Neil Gaiman unpacks the role of interpretation in representation:

> One describes a tale best by telling the tale. You see? The way one describes a story, to oneself or to the world, is by telling the story. It is a balancing act and it is a dream. The more accurate the map, the more it resembles the territory. The most accurate map possible would be the territory, and thus would be perfectly accurate and perfectly useless. The tale is the map which is the territory. You must remember this.
>
> (Gaiman 2006: 10)

What Gaiman succinctly articulates is how representation is impossible without meaningful alteration. Traditional scholarship strives to produce the most faithful possible representation of a text. The most accurate representation, however, is no longer a representation; it is an identical reproduction. To engage with a text in a meaningful way requires an interpretative retelling of the text's story. In Hayden White's words, '[n]arrative . . . entails ontological and epistemic choices with distinct ideological and even specifically political implications' (1987: ix). Reading, in this sense, is not a neutral unveiling of the text's 'true' meaning – it is a political act in which every reader emphasises certain aspects, marginalises others and excludes some things altogether: it is to describe a story by telling it. In this respect, writing can never escape the possibility of non-transcendent reading; and the Derridean understanding of reading allows us to re-frame the relationship between interpretation and reading – not as the fall from rigour or authenticity as traditional ideals of neutrality imply, but as the positive possibility of engaging with any text. A scholarly retelling that doesn't involve some element of interpretation necessarily takes the form of an identical reproduction. To engage with any text in a meaningful way therefore necessitates interpretation.

Derrida maintains that to read a text is to reconfigure it in some meaningful way. Zehfuss explains that non-transcendent reading is not concerned with the illusionary ideal of finding the intended meaning; instead, the goal is to tease out the multiple interpretations and implications that can be found in a text (2007: 23). What is important is Zehfuss's contention that every text contains multiple significations: there are multiple ways in which the story of a text can be told, i.e. texts do and say things other than what their authors intended. Again, what is paramount is the negotiation between conservation and alteration. The reading of Walzer presented throughout this book does not attempt to misrepresent him or attribute to him things that he has not said. Nor does it seek to protect a particular reading of Walzer that reinforces the overarching conception of responsibility he provides. Instead, the following chapters seek to demonstrate, through an attentive reading of Walzer's work, that the arguments he wants to make – the models of justice, community and war that he presents us with – leave him with no option but to risk saying something other than what he intends to say. As such, the purpose of this work is partly to illustrate why Walzer's conception of justified violence collapses under the weight of its own tensions, inconsistencies and assumptions. However, this reading does not aim to completely renounce Walzer's understanding of ethics. Rather, it suggests that challenging Walzer's most pervasive assumptions and beliefs about violence, politics and justice can help us formulate new understandings and new answers to the ethical questions posed by war.

Historical analysis and the problem of representation

This book attempts to situate the critique of Walzer within the context of the Iraq War; historical analysis is thus a key methodological component of how this book functions. Nevertheless, the use of history presents a number of problems for those of us drawing upon post-structural philosophy. When discussing contemporary approaches to historical analysis, Nick Vaughan-Williams asks a pivotal question: 'What do we mean when we refer to *History* in IR?'(2005: 115). To this end, Vaughan-Williams argues that IR scholars have failed to reflect upon this question and have, in general, accepted the view that history is a quarry to be mined in support of theories in the present. In other words, IR scholars have viewed history as a form of unproblematic empirical evidence that they can use to support their theoretical arguments. In response to this understanding of history, Vaughan-Williams called for IR scholarship to embrace the 'problem of history' and take the interpretative dimensions of historical analysis into account (ibid.). Broadly speaking, critical approaches to IR emphasise the pivotal role of interpretation in our engagement with the world.[5] Pin-Fat (2005), for example, argues that our understandings and explanations of the world around us derive in part from the metaphysical assumptions we already hold; that is, we interpret the world through pre-existing frameworks of beliefs and assumptions – and this impacts upon how we describe events and phenomena in the international realm. Given the seeming widespread acceptance, among critical IR scholars, that our interpretation of the world impacts upon our understandings, explanations and analysis of it, the surprise therefore lies in how few scholars have attempted to broach the question of the relationship between critical methodologies and historical analysis. This is perhaps a result of a tacit assumption that the interpretative nature of truth, in general, necessarily correlates to an interpretative understanding of history. For example, when Steve Smith argues that all perceived empirical truths are 'a matter for negotiation and interpretation', we can assume that this encompasses historical accounts too (2004: 514). Nevertheless, the assumption that historical accounts are interpretative and contested fails to account for the implications of this assumption and the difficult questions it raises about the uses of history within critical scholarship.

These questions are particularly pertinent for those of us writing on the topic of war. One of the fundamental questions faced by scholars engaged in war studies is: how do we find out about what happens during war? The fact that academics rarely have direct access to arenas of war (owing to security concerns, travel restrictions, and so on) makes this a pressing concern. The staple response to this problem is that we seek out accounts from those who have witnessed the events, or turn to existing literature that has already compiled such accounts. War studies, therefore, relies upon historical accounts; it relies upon the information we garner from journalists, combatants and other witnesses. This may all seem rather straightforward; we seek out relevant information as a means to help us explain and understand events and phenomena that we cannot personally investigate. However, for critical scholars, this raises some difficult methodological questions. Specifically, it highlights the spectre of how critical methodologies sit uneasily with the commonly perceived evidential quality attached to first- and second-hand historical accounts.

Vaughan-Williams claims that the traditional understanding of historical analysis in IR sees theory and history as two separate terrains: conceptual arguments and the empirical evidence we use to support these arguments (2005: 134). This conception of historical analysis follows a similar trajectory to what Carroll Quigley describes as the 'scientific method' of historical analysis. Quigley argues that when we engage with historical sources we must follow a three-step methodology, in which we gather relevant evidence, make a hypothesis and test the hypothesis (1979: 33). In Quigley's model, history serves as the relevant evidence against which we test our theoretical hypothesis. History, in short, represents the *proof.* If we find enough proof to substantiate the hypothesis, then we have demonstrated the saliency of our claims.[6] Walzer embraces this ideal of history as demonstrable proof. He makes a direct link between drawing upon historical sources and accurately depicting reality: 'Since I am concerned with actual judgements and justifications, I shall turn regularly to historical cases' (2006a: xxiv). Walzer expands on this motif by explaining how historical examples help him to illuminate the moral reality of war (2006a: 14–15). In other words, Walzer uses history as evidence to illustrate why the philosophical arguments he is making are directly related and applicable to the real world. He presents historical examples as the 'hard evidence' needed to demonstrate how his arguments reveal the moral reality of war. Like Quigley's scientific method, historical sources are deployed as evidence to validate and verify theoretical claims: Walzer outlines a particular moral argument (theory) which is, in turn, demonstrated by presenting us with a corresponding historical example (empirical evidence). Walzer's understanding of history, however, introduces an additional dimension to the treatment of historical accounts in war studies: the correlation between first-hand experiences of war and authoritative truth claims. Walzer argues that drawing upon first-hand accounts adds weight to his moral arguments. He implies that because he is 'reporting on experiences that men and women have really had and arguments that they have really made', his moral arguments are imbued with a sense of palpable authenticity (2006a: xxiv).

This project wants to tentatively embrace history while simultaneously challenging Walzer's understanding of historical analysis. More specifically, I want to retain Walzer's engagement with the real world consequences of war while taking account of the interpretative dimensions of historical analysis. As outlined above, academics in war studies are rarely afforded first-hand experience of the conflicts we investigate. It is therefore unsurprising that we often feel inclined to accede to the authority of those who have lived through war. After all, these are people who have braved the danger and suffered the costs of violence. This acceptance of the authoritative voice of the first-hand witness is highly problematic, however, and ignores the implications of interpretation discussed in the previous section. Primarily, it fails to account for the socio-political contexts in which the events of particular wars are witnessed, or the existing beliefs and understandings those recounting events bring to their recollections. The ideal that historical accounts serve as a form of proof is linked to classical anthropology's suggestion that first-hand exposure opens the possibility of a faithful representation. Claude Lévi-Strauss, for instance, defines the methodology of classical ethnography as the aim to record as accurately as possible the respective mode of life of various groups through first-hand exposure (1974: 2). This ideal positively correlates first-hand experience with the observer's capacity to provide an accurate account. An observer with first-hand experience of an event or phenomenon, therefore, is in the best position to uncover the truth. Derrida (and others, including Lévi-Strauss) reject this conception of representation because it excludes the implications of the broader social contexts that underpin our engagements with the world. The approach to history outlined in this book therefore takes into account the complex relationship between interpretation and representation in the production of first-hand accounts of war. In short, it asks us to view the history of various wars as texts.

It is useful to view history as text for a number of reasons. In the first instance, those narrating the events of war already have a pre-existing system of beliefs: they view the event from a singular perspective with distinct cultural and social markers and belief systems. In *Of Grammatology*, Derrida provocatively declared that '[*T*]*here is nothing outside of the text*' (1997: 158, original italics). Derrida's provocation is often read as a denial that material reality exists outside language and as a privileging of the ideational world over the world of objects.[7] However, Derrida's argument is far more concerned with the idea of mediated engagement: it is not that the material, or real, world does not exist abstracted from language; rather, interaction with material reality is predicated upon a cognitive apparatus that necessitates language. In other words, language is necessary to render the world of objects meaningful – and therefore, the possibility of engagement with the real world is underpinned by the existence of language. Derrida clarifies his argument in the afterword to *Limited Inc*, explaining that the term 'text' can be re-read as context, and context should be understood as the entire 'real-history-of-the-world' (1988: 136–137). What Derrida wants to emphasise is that the contexts through which we begin to engage with the world have important implications for how we interpret and make sense of our realities. In this sense, to refer to war as a text implies that our engagement with war is mediated through the contexts in which we encounter it. The contexts surrounding our engagements with war, in important respects, are implicated in the ways in which we interpret the events and the stories we tell about them: our understandings are dependent upon who we are (the prior assumptions and under-standings that we bring into our engagement), why we are engaging with this specific conflict and the media from which we garner information about the conflict.

The widely reported toppling of a statue of Saddam Hussein in the aftermath of the Iraq War in 2003 provides a good example of the implications of this. The toppling of the statue was a pivotal moment in the US imagining of Iraq, with US media outlets portraying the event in terms of symbolic unity between Iraqis and coalition troops against Ba'athist rule.[8] The common media narrative claimed that a group of Iraqis were chipping away at the base of the statue for hours to no avail. Witnessing the failed attempts, US Marines stepped in, dragging the statue to the floor with their tanks. Iraqis were said to have responded in jubilant triumph, slapping the statue's decapitated head (*New York Times*, 9 April 2003). In many respects, this representation of the event echoed the overarching Bush Administration's justification of the war: US military might helping Iraqis to do what they wanted, but lacked the power, to do. Yet reactions within the Baghdad crowd comprised a more fragmented and confused response than what the US media portrayed. The crowd's behaviour, for instance, was dominated by feelings of anger toward Ba'athism rather than joy at liberation, and an undercurrent of humiliation emerged when Marines draped the statue in a US flag (BBC News, 9 April 2003). Importantly, the primarily Shi'a crowd vocally expressed their anger, joy and sorrow in explicitly religious terms. In the words of the late journalist Anthony Shadid,

> As they toiled, groups of religious Shiite Muslims gathered on the side of the square. 'There is no god but God; Saddam is the enemy of God,' they chanted. Some seized the opportunity to pray in open, no longer at risk of suspicious stares. Others beat their chests in a ritual of grief known as *lutm*. It was the first time I had seen such a display of religious practice in public in Iraq.
>
> (Shadid 2006: 148, original italics)

What is important in the various accounts is that no specific perspective provides us with a more authentic version of the event; rather, it underscores the fact that the perspective from where the witness is coming radically transforms the type of narrative they tell. Shadid's narrative, for example, emphasises how the reactions of Iraqis to the event highlighted a cultural shift; whereas the *New York Times* focuses more directly on the US Marines and the symbolism of the event in terms of the US war effort. The composition of first-hand accounts of events in war, therefore, contains an implicit interpretative dimension.

First-hand accounts, in this sense, are contextually dependent interpretations of events rather than cold hard facts. The journalists, soldiers, politicians, academics and others who have relayed stories from war to those of us outside the conflict zone are presenting narratives based on the contexts of their engagements: who they are, where they come from, why they are telling the story, and so on. In other words, the possibility of telling any story about war is underpinned by the narrator's interpretation of their limited experiences of the war, and the ways in which they reconstruct this interpretation into a communicable narrative. As Shadid maintains, no account can convey the complex intricacies and confusion of war; the only thing witnesses can do is tell stories (2006: 12). As such, relaying the history of war is another instance of textual interpretation. Specific wars are texts that are interpreted by witnesses in various ways and subsequently transformed into narratives that are relayed to, and interpreted by, others.

The history of war, therefore, moves between a multiplicity of authors and readers within a textual structure. To expand upon this motif, let us look at how first-hand journalistic accounts of the Iraq War were produced. Journalists who reported on Iraq can broadly be divided into two distinct groups: those embedded with US troops, like Evan Wright and David Finkel; and those who operated independently, like Shadid and Dahar Jamail. Both of these avenues restricted reporters' access to the Iraqi text. On one hand, embedded reporters (as well as combatant authors) could only travel with troops and were often obliged to submit their work for military approval before publication. While on the other hand, the movement of independent reporters was often restricted due to security concerns and they were ultimately dependent on the testimony of locals, who interpreted events in their own ways. Narratives about the Iraq War thus represent a limited, contextually dependent, interpretation of a much bigger picture. Even when we combine accounts, as the majority of work in war studies does, we are unable to circumvent the problems of context and representation. Combined accounts lead us back to the role of the scholar who merely provides a collage based upon the particular context of their engagement and how they interpret their sources. First-hand narratives, therefore, cannot unveil the truth of war. They can only provide us with contextually restricted interpretations of the truth. This is equally true for historical accounts of other forms of events. This is what Vaughan-Williams calls the 'problem of history': historical events can never exist in the form of a singular reality because we can only access them through a multiplicity of interpretative accounts.

With whom does the historical writer of historicism actually empathise?

The last section underscored the 'problem of history' and questioned our ability to use history as a form of demonstrative evidence. I argued that the interpretative dimensions implicated within every possible engagement with the world around us deny the possibility of using historical accounts as a source of empirical truth. Instead, history provides a myriad of – sometimes complementary and sometimes competing – interpretations of the truth. Despite the problems inherent to any account of history, Derridean thought reminds us that this is the only means we have of engaging with the world – that is, if we want to garner any understanding of history, or any other topic for that matter, we have no option but to engage with contextually limited narratives. Walter Benjamin's (1999) understanding of historical analysis provides an interesting way to think about how we can critically engage with the history of the Iraq War. Benjamin's 'Thesis on the Philosophy of History' attempts to re-frame historical enquiry in terms of political engagement. Benjamin contends that history is a ground of political contestation, an arena in which multiple interpretations are possible and can be used to emphasise contrasting political objectives. David Campbell's (1998a) reading of the Bosnian conflict, for example, emphasises the contemporary dimensions of the civil war; whereas Robert Kaplan (1993) describes the animosities that arose during the 1990s as deep-rooted historical antagonisms. In this sense, drawing upon historical sources is not a neutral act of recollection. It is a political act of interpretation. To recall history is to present a particular interpretation of events that are linked to particular understandings of the world and particular ethico-political motivations.

Central to Benjamin's conception of history maintains that reading historical sources implies engagement with a past that is never settled as historical truth. For Benjamin, there is no singular historical truth; rather, there is a multiplicity of potential interpretations that can be mobilised to achieve different aims. Benjamin argues that the interpretation of history is a political tool that can help us respond to present concerns (1999: 247). When military intervention is presented as a possible response to the repression of a set of people, for instance, we turn to historical cases to help us construct our arguments for or against the proposed military action. In this regard, Benjamin argues that history is infused with what he calls 'the presence of the now' (1999: 253). What this implies is that the political power of history does not reside in the traditional focus on recalling the past as past, but in the critical desire to politically mobilise an interpretation of the past in a way that relates to a present context. For example, arguments in favour of intervention in Syria in 2013 suggested that lessons we learned from the Iraq War would ensure that we would not repeat the same mistakes.[9] In turn, Benjamin contends that historical analysis has revolutionary potential, because we can interpret history in a way that helps produce the possibility of new socio-political formations. Benjamin's conception of historical analysis as a revolutionary opportunity is aligned with the Derrida concept of non-transcendent reading. In both instances, we look toward the interpretative dimensions of scholarship as a means to open new possibilities in the present. Nevertheless, Benjamin does not forget that the political dimension of historical analysis also means that history can be interpreted in a way that preserves existing socio-political configurations and closes possibilities (1999: 247). In this respect, Benjamin points toward the explicitly political character of historical analysis: history is the battleground for contemporary political arguments in which various historical narratives are mobilised and put to work within broader political agendas.

Central to Benjamin's understanding of historical analysis is the concept of 'monad'. Benjamin describes monads as aspects of history that can be redeployed as a revolutionary opportunity to put history to work in present day politics:

> Where thinking suddenly stops in a configuration pregnant with tensions, it gives that configuration a shock, by which it crystallises into a monad. A historical materialist approaches a historical subject only where he encounters it as a monad. In this structure he recognises the sign of a Messianic cessation of happening, or, put differently, a revolutionary chance in the fight for the oppressed past.
>
> (Benjamin 1999: 254)

Monads, in short, are instances in which past oppressions are reflected within present political struggles. If we unpack monads a bit, they help us to explain what critical scholars do when they engage with history. Benjamin claims that history is not encountered in the form of a resolved empirical truth. Instead, history is approached in an interpretative process through which particular readings aim to bring about particular changes in the present. Importantly, history is itself uncertain and 'pregnant with tensions' that create the possibility of diverse interpretations. In other words, historical narratives remain textual constructs that contain the possibility of preserving or destabilising particular meanings and understandings. Thus, narratives of particular historical events contain within the structure of their texts either the possibility of conserving accepted understandings or opening the way for new understandings and explanations. Historical events are not resolved as singular truths; as such, the lessons we have supposedly already learned from them are open to critique and reformulation. This has important ramifications for the study of war. Reading the history of specific wars constitutes a political act: different readings of historical events can encourage different responses to the questions posed by war and violence. Crucially, reading history in a critical manner can open the way to alternative understandings and explanations of war.

Presented with this explicitly political understanding of historical analysis, the primary question is no longer about how we engage with historical sources – instead, what we need to ask ourselves is: why do we engage with them? What motivates our research? What do we want it to accomplish? And how are the ways in which we interpret and unpack accounts of warfare related to these aims and motivations? In his thesis, Benjamin (1999: 248) asks an important rhetorical question: with whom does the historian actually empathise? Benjamin answers that historians have traditionally empathised with the 'victor'. He argues that historians are primarily heirs to the victors' civilisations and historical analysis has thus been largely in the service of preserving existing orders. In juxtaposition, Benjamin contends that to approach history in a critical manner requires us to interpret history in a way that empathises with the 'oppressed'.

Benjamin's twin allusion to empathy and past oppression offers what I think is an important acknowledgement in relation to the question of why IR scholars engage with the subject of war. Empathy is, in certain respects, an important facet of our engagement with war. Prominent here is my belief that most people who begin to think about the subject are to some extent inspired to do so because they are aware of the devastating consequences war has inflicted upon people in the past. In this regard, our exposure to the historical consequences of war on other people stimulates a desire to reduce suffering in the present day; our exposure to the horror of war provokes the call for a response. To generate empathy for those oppressed by war is, therefore, an important political aspect of critical approaches to war studies. This is, perhaps, the antithesis of the proposed detached and neutral stance endorsed in scientific and certain analytical methods. The appeal to empathy constitutes an ethico-political position that directly relates scholars to people affected by the consequences of war. It no longer views war as an object of study, but as a context in which thousands of subjects suffer on a daily basis. It is within this strategy of fostering empathy that historical accounts of war hold much promise. Our empathies are stimulated, not by abstract analogies or thought experiments in which nothing is really at stake, but by actual tragedies of war and the people they affect. Those involved in war are, in Benjamin's terms, oppressed peoples facing impossible conditions that generate complex political and ethical questions that demand a response. Our responses to these questions need to be honest in their political positions and aims. Critical scholars cannot simply present interpretations of history as a form of apolitical evidence. Instead, we need to take ownership and accept accountability for the political arguments we are making, and firmly state why we adopt particular political platforms and to what end. Ultimately, it is only through clearly demarking the ethico-political nature of academic debates in IR that we can re-politicise some of the most important assumptions that traditional approaches seemingly take for granted.

This work reads the history of Iraq in terms of empathetic solidarity with the oppressed: it engages with historical narratives as a means to illustrate the ways in which conventional ideas of ethical responsibility in war have contributed to the oppression of a myriad of different peoples. In this respect, the Iraq War is interpreted in a way that emphasises the inadequacy of Walzer's account of justice, and draws our attention to possibilities offered by Derridean thought. The method of historical engagement outlined here does not correspond to a linear relationship between theoretical argument and historical evidence. It does not suggest that the reading offered here, or anywhere else, can be definitive, conclusive or wholly faithful to the subject of discussion. In contrast, this book is an ethico-political strategy designed to impress the need for alternative understandings of what ethical responsibility entails in the context of war through a particular, albeit careful, reading of numerous accounts of the invasion of Iraq and its aftermath. Historical analysis is, in this way, re-joined with the deconstructive method: I offer a respectful reading of first-hand accounts of the Iraq War that is, nonetheless, strategically focused upon opening discourses concerned with the ethical consequences of war to the possibility that alternative accounts of ethical responsibility can, perhaps, reduce the suffering inflicted upon those oppressed by war.

Structure of the book

This book presents an alternative conception of ethical responsibility through a critical reading of Michael Walzer's just war theory and a historical analysis of the US invasion and subsequent occupation of Iraq. To achieve this, the book follows a clear structure of exposition. First, it provides a critical reading of Walzer's theory of war and understanding of ethical responsibility. Second, it provides some historical background to Iraq and explains how Walzer's conception of ethics is linked to US actions in Iraq. Third, I outline an alternative conception of ethical responsibility – ethics as response – which challenges Walzer's understanding of morality and the justifications given for the invasion of Iraq. Finally, the book provides a critical analysis of the two pillars of Walzer's theory of war, non-combatant immunity and double effect, illustrating how ethics as response can help us to better understand the ethical issues raised during the Iraq War.

The following chapter provides a critical examination of Walzer's conception of wartime morality. The central aim of this chapter is to illustrate why Walzer's justification of war fails to function in the way that he wants it to. I contend that Walzer justifies war because he believes that violence is necessary to defend communal self-determination. As such, Walzer's justification of war is intimately related to his broader communitarian project. The main argument presented in this chapter is that Walzer is unable to justify war as a defence of self-determination because this justification is founded upon inadequate accounts of community and morality. The second chapter provides a brief historical analysis of Iraq. The chapter begins by outlining Iraq's pre-war society and politics. However, the central focus is on explaining how the 2003 invasion and the post-war US occupation of Iraq radically re-shaped Iraqi society. In this respect, the chapter will focus on the relationship between the Bush Administration's justification of the war and Walzer's theoretical arguments. In turn, this chapter seeks to illustrate the social and political consequences of violence and, in particular, how intervention can redefine communities and their internal dynamics.

Chapter 3 unpacks an understanding of ethical responsibility that challenges Walzer's justification of war, and the Bush Administration's conception of justice in post-war Iraq: ethics as response. The primary aim of this chapter is to highlight why our desire to help others, and to respond toward other people, necessitates a simultaneous affirmation and sacrifice of ethical responsibility. The central argument posited in this chapter is that war cannot be justified, in the way Walzer suggests, because it must risk unintended and unforeseeable consequences. In this respect, ethical responsibility is repositioned as a negotiation between our desire to help people and our acknowledgement that war necessarily risks adversely effecting or affecting others; a negotiation between responsibility and irresponsibility.

The fourth chapter focuses on Walzer's justification for the killing of combatants. Walzer declares that the right of combatants to kill or be killed is the foundation of his rules of war (2006a: 136) – so intrinsically, the possibility of justifying any act of war depends upon this rationalisation. The main argument presented here is that Walzer's conception of non-combatant immunity fails on its own terms and, more importantly, that it is unable to account for the ethical implications of sending men and women into an arena in which they are expected to kill or be killed. Through a discussion on Walzer's theoretical arguments, and a reading of US combatant narratives from Iraq, this chapter illustrates why the experiences of US combatants in Iraq confirms the Derridean account of ethical responsibility; Walzer's conception of just war is possible only by sacrificing our responsibilities toward combatants.

Chapter 5 engages with the second pillar of Walzer's rules of war: the doctrine of double effect. Walzer argues that double effect justifies the 'unintentional' killing of non-combatants during war. He contends that the killing of civilians is justified so long as it is an unintentional consequence of a legitimate act of war. This chapter argues that Walzer is unable to sustain his conception of intentionality because it splits responsibility from its unintended consequences. Drawing upon ethics as response, I contend that the risk of unintended and unforeseeable consequences is integral to the possibility of any action and, accordingly, Walzer cannot exclude unintentional effects as a means to justify the killing of civilians. This chapter concludes by suggesting that the decision to go to war is itself an instance of double effect because it risks changing the socio-political landscape in unintended and unforeseen ways.

Conclusion

The underlying argument presented in this book is that questions of ethical responsibility in war remain irresolvable. This represents an important departure from the types of arguments traditionally associated with the JWT and general discussions on ethics in war. Traditional discussions have been largely focused upon designing general rules, laws and norms that contend to solve the questions of when war is morally justified and what it means to act responsibly in war. In other words, the goal of traditional discussions is to tell us what is morally right and morally wrong when it comes to war – and Walzer's work provides an influential example of this form of argument. While these discussions remain an important aspect of our understandings of the relationship between ethical responsibility and violence, they nevertheless endorse an ideal of ethico-political and critical disengagement. Attempts to solve the questions raised by war ask us to situate our thinking within particular frameworks and, more importantly, they equate responsibility with acting in accordance to moral rules, laws and norms. Acting responsibly thus means following the moral law rather than responding to the needs and demands of other people. Ethics as response is an attempt to explain what it means to ethically respond to other people in the absence of definitive rules and norms. As such, it is an attempt to describe an understanding of ethical responsibility and action in which we are never certain what is the right or wrong thing to do. Importantly, ethics as response is a call for us to remain critically invested in the difficult questions raised by the relationship between ethics, politics and violence, without the possibility of resolution or cessation. It is a conception of ethical responsibility that refuses to satisfy itself in the adherence to rules and instead, demands that we stay involved, and remain involved, in the consequences and new contexts that our responses help to create.

Notes

1 I employ the term US rather than America to denote the United States of America where possible. However, the terms America and American are used when either directly quoted by other authors, or if they are directly related to concepts that other authors employ.
2 See Kenneth Roth, 'War in Iraq: Not a Humanitarian Intervention', in Richard Ashby Willson (2005) *Human Rights in the 'War on Terror'* (Cambridge: Cambridge University Press).
3 For example, see Alex Bellamy (2008), *Responsibility to Protect* (Cambridge: Polity Press), J. L. Holzgrefe and Robert O' Keohane (2003), *Humanitarian Intervention: Ethical, Legal and Political Dilemmas* (Cambridge: Cambridge University Press), Mary Kaldor (1999), *New and Old Wars: Organised Violence in the Global Era* (Cambridge: Polity Press), James Pattison (2012), *Humanitarian Intervention and the Responsibility to Protect: Who Should Intervene?* (Oxford: Oxford University Press), Thomas Weiss (2012), *Humanitarian Intervention (War and Conflict in the Modern World)* (Cambridge: Polity Press) and Nicholas J. Wheeler (2000), *Saving Strangers: Humanitarian Intervention in International Society* (Oxford: Oxford University Press).
4 In this respect, Walzer's argument is not entirely distinct from Kantian liberalism. For example, see Andrew Linklater (1982), *Men and Citizens in the Theory of International Relations* (London: Palgrave Macmillan).
5 For example, see Richard K. Ashley (1984), 'The Poverty of Neo-Realism', *International Organisation*, 38(2), pp.225–286, Véronique Pin-Fat (2010), *Universality, Ethics and International Relations: A Grammatical Reading* (London: Routledge) and Steve Smith (2004), 'Singing Our World into Existence: International Relations Theory and September 11', *International Studies Quarterly*, 48(3), pp.499–515.
6 It must be acknowledged that Quigley maintains that no proof is definitive and remains open to contestation.
7 For example, see Perry Anderson (1983), *In the Tracks of Historical Materialism* (London: Verso).
8 For example, see CNN, 'Saddam Statue Toppled in Central Baghdad', (9 April 2003), http://edition.cnn.com/2003/WORLD/meast/04/09/sprj.irq.statue/
9 See Daniel Politi, 'Obama: Syria "Would not be another Iraq or Afghanistan"', *Slate*, 7 September 2013, http://www.slate.com/blogs/the_slatest/2013/09/07/barack_obama_ weekly_address_president_makes_case_for_syria_strike_ahead.html

1 Morality, community and the justification of war

Introduction

In an interview with *The Art of Theory* in 2012, Michael Walzer claimed that the biggest success of his theory of war was its adoption as a text by the West Point Military Academy during the Vietnam War. According to Walzer, while service-men may not have agreed with his views on the war, their discomfort with the war had made them receptive to his moral arguments. Walzer, in this respect, wants his arguments to have practical impact: he wants publics to engage with them and, particularly for those involved in fighting wars, to be challenged by moral dimensions of contemporary conflict. This is, undoubtedly, an important and noble aim. Indeed, provoking people to think critically about war is perhaps the overarching theme that ties together the vast majority of academic literature on ethics and war. Nevertheless, Walzer's desire to achieve practical impacts also carries some weighty risks: principally, the risk that his moral defence of war will be deployed as a strategic tool to justify violence. In other words, Walzer's pride at his work's military adoption should be, perhaps, tempered by a sense of unease: the nagging worry that his justifications can be employed in the service of immoral goals. Given the risks associated with justifying war in moral terms, it is crucial for us to understand what morality means to Walzer and how this relates to his justification of war. Therefore, the central focus of this chapter is to explain Walzer's justification of war by unpacking his understanding of morality. To this end I will be critically engaging with Walzer's depictions of morality, community and ethical responsibility. In addressing these themes, this chapter will provide an account of Walzer's overarching philosophical and political project, thereby providing a platform to elucidate an alternative understanding of what it means to act ethically in times of war.

This book describes Walzer's conception of ethical responsibility as *auto-affective*. This means that Walzer believes that ethics can only emerge from within the stable boundaries of self-determining subjects. Hence, the way in which we relate to other people is predicated upon a movement from self to other that begins with the self: we start with separated and internally coherent subjects who can subsequently engage with their outside. In turn, this model of ethical responsi-bility is mirrored in Walzer's depiction of morality. For Walzer, we start with morality produced within particular communities and this creates the possibility of producing inter-communal, sometimes universal, codes. In this sense, we start with separated, internally coherent communal subjects and this opens the possibility of ethical engagement between communities. This understanding of ethics is important to Walzer's justification of war because he maintains that violence is justified only when it is necessary to protect the self-determining communal subject. In this sense, this chapter highlights why it is important to understand Walzer's theory of war within the context of his wider communitarian writings. By situating Walzer's just war theory within the communitarian arc, we can more clearly see his justification of war in terms of a defence of communitarianism.

The challenge to Walzer's auto-affective ethics presented in this book arises from a Derridean understanding of the relationship between meaning, subjectivity and responsibility. This model of ethics will be more clearly articulated in the third chapter. Nevertheless, to understand the critique presented within this chapter, it is important to explain the Derridean concept of *the law of supplementary commencement*. Derrida (1997) argues that when we attempt to locate a singular and definitive ontological origin, what we actually find is a chain of supplementary origins: what Derrida describes as a *non-origin*. The supplementary non-origin indicates that the foundation we hope to locate was always already in motion and, therefore, there is never a clear starting point through which we can ground ontology. As will be illustrated throughout this chapter, supplementary commencement challenges auto-affection by highlighting the relational dimensions implicated in the emergence of a coherent inside, in particular the constitutive role of alterity in the production of the communal subject. Walzer's model starts with the assumption that self-determining, internally coherent subjects exist prior to their relationship with the outside – prior to ethical relationships. The Derridean critique, however, explains why the relationship with alterity is constitutive of subjectivity itself. In this sense, the self-productive model of subjectivity, community and meaning implicit to Walzer's argument constitutes an inadequate ontological foundation.

Walzer, as discussed in the introduction to this book, grounds his morality in the language of rights. The language of rights constitutes a universal structure that allows Walzer to theorise morality in war – morality shared across all communities. In addition, Walzer considers this language to provide morality with a secular ontological foundation. Nevertheless, this ontological foundation contains important theological components. In destabilising the foundations of Walzer's ontological system, and paying close attention to the necessity of alterity in its production, we can grasp the theological dimension necessarily retained within this supposedly secular morality. I contend that the universal morality necessary for the foundation of Walzer's laws of war embodies a form of what I term 'secular theology'.[1] By this I mean that Walzer's ontology is punctuated by unacknowledged transcendental appeals to faith, without which he would be wholly unable to establish or sustain his system of morality. Reading Walzer's morality in terms of secular theology allows us to re-present Walzer's 'moral reality of war' as the unfounded imposition of a particular interpretation of morality as ontological fact. In this way, Walzer's wartime morality is reconceptualised as a socio-political strategy designed to protect a particular understanding of self-determination intimately related to a particular understanding of ethics.

There is a thin man inside every fat man

Walzer's seminal work on war, *Just and Unjust Wars*, was primarily a response to what he perceived to be an ethical debasement of the subject spearheaded by realist thinkers. What is perhaps most interesting about Walzer's response is that it challenged realism on its own terms. Forgoing the traditional liberal stance that morality was something that needed to be worked into the mechanics of war, Walzer argued that morality was already, and always had been, a tangible component of the reality of warfare. In this way, Walzer challenged realism, not with what could simply be dismissed as moral naivety or good intentions, but with reality itself, claiming that the reality espoused by realism constituted a crude fiction used to justify immoral actions: 'we do not have to translate moral talk into interest talk in order to understand it; morality refers in its own way to the real world' (2006a: 14). In contrast to the deceptive language of realism, Walzer describes the language of just war theory, at various junctures, as: the ordinary language of war (2005: 8); a common heritage (2005: xi); the most available common moral language (2005: 7); and a moral doctrine that everyone knows (2006a: xix). Walzer's underlying argument is that when we discuss the issue of war, we 'talk the same language', the language of just war – and only the wicked or the simple would reject its terms (2006a: xxiii). In this respect, although Walzer states his intention to defend the business of arguing about war, he quite literally wants to fix the terms of this debate: 'it is in applying the agreed-upon terms to actual cases that we come to disagree' (2006a: 11–12) – that is, Walzer presents us with the necessity for an agreed-upon common language that allows us to critically engage with the moral reality of war, and this language is embodied by the terminology of just war theory. Fundamentally, Walzer is proposing a counter-ontology to realism's language of power and interest, in which just war's vocabulary allows us to illuminate the moral reality of war.

While Walzer's depiction of a shared moral language may seem relatively straightforward, its articulation proves more complex than it initially appears. The fundamental complication within Walzer's understanding of morality derives from his claim that there are two distinct, but not mutually exclusive, languages of morality: what Walzer terms thick and thin moralities. For Walzer, this dichotomy represents a dual affirmation of particularism and universalism, a politics of difference coupled with the acknowledgement of universal rights (1994: x). Thick, or maximal, moral language is described as the shared meanings of a singular political community, representing its collective conscience and common life: Walzer's conception of particularism (1994: 8). Morality is negotiated thickly between members of a community, ultimately creating a common social vocabulary. Through this shared vocabulary, members define their laws, ideals, values and institutions. There are important ontological problems with Walzer's conception of community, problems that will be addressed in the following sections.

Nevertheless, this does not necessarily concern the laws of war because thick morality cannot be universalised. Walzer assures us that the authority of maximal morality is rooted in the singular community and any attempt to enforce thick standards in another community (by an outside party) violates that community's right to territorial integrity, political sovereignty and self-determination (2006a: 53–55, 61).[2] Because Walzer's rules of war are designed to be enforced across communities, we must turn our attention to the language of thin, or minimal, morality, the universal moral vocabulary and therefore, the universal dialect of wartime morality.

Walzer quickly asserts that minimalism is best understood as an effort to recognise and respect a doctrine of rights (2006a: xxiii–xxiv). While Walzer is unsure where rights derive from (whether they are natural or invented), he assures us that they are inseparable from our sense of what it means to be human and constitute a palpable feature of our moral world (2006a: 54). In turn, although Walzer recognises that rights are a form of Western maximal language, he assumes they are translatable (1994: 10). In his conceptualisation, Walzer describes the rights of life and liberty as something more than simply minimal – what he terms 'ultra-minimalism' (1994: 16). In this sense, the rights of life and liberty represent the core minimal essence of universal morality. In the context of war, Walzer asserts that the rights of life and liberty 'underlie the most important judgements that we can make about war' (2006a: 54) and we can only justly send armed men and women across a border in defence of life and liberty (1994: 16). Importantly, Walzer argues that justice in war can be derived exclusively from the protection of life and liberty: 'For the theory of justice in war can indeed be generated from the two most basic and widely recognised rights of human beings – and in their simplest (negative) form not to be robbed of life and liberty' (1983: xv). The rights of life and liberty are, in this regard, the foundational components of Walzer's universalism. In fact, Walzer maintains that life and liberty should be viewed as absolute values that dictate every moral judgement we make at times of war (2006a: xxiv). Absolute, for Walzer, is best understood in terms of inviolability: life and liberty are rights that cannot be violated without acting immorally. In this context, the War Convention (Walzer's codification of the moral rules of war) is underpinned by the defence of life and liberty and is therefore 'written in absolutist terms: one violates its provisions at one's moral, as at one's physical peril' (2006a: 47).

It must be underscored that Walzer's conception of rights is not equivalent to that espoused by classical rights theorists, as exemplified by the Rawlsian (1999) model that presents individual rights as the foundation of universal morality and justice. In contrast, Walzer argues that rights emerge from the jagged bedrock of particularism: we start with communal systems of morality that coalesce into universal rights. In this way, the codes of maximal morality produced within individual communities provide the foundation through which the universal laws of war can emerge. As Walzer explains, '[m]orality is thick from the beginning, culturally integrated, fully resonant, and it reveals itself thinly only on special occasions' (1994: 4). In other words, we start with particular codes of morality and these facilitate the emergence of universalism. Here, Walzer performs a clever linguistic trick: rather than offering a singular universal morality, he creates the image of numerous and diverse maximal moralities dovetailing into a set of universal guidelines. Walzer claims that minimal morality represents a catalogue of common responses that combine to form a set of standards to which all societies can be held (1994: 10). As such, Walzer's universalism resembles the intersection of a vast inter-communal Venn diagram, symbolising a negotiated minimal code rather than the enforcement of a singular set of universal values.

The intersecting point of Walzer's moral diagram captures the minimal essence of life and liberty and the moral rules constructed to protect these absolute rights. This image mirrors Walzer's depiction of the War Convention, which he assures us is the product of centuries of inter-communal arguing and debate over the morality of warfare (2006a: 44–45). Walzer, in this way, illustrates his interpretation of the Orwellian metaphor of the thin man inside the fat man: thin morality emerges from thick moralities; universalism is founded by particularism. This conception of morality tells us a great deal about Walzer's view of ethics. Ethical responsibility begins with the coherent communal subject and responsibility is defined in terms of the relations between members – maximal morality. In turn, the existence of a coherent and stable inside makes the inter-communal rules of war, or minimal morality, possible, i.e. ethical responsibility starts in the self (the communal subject) and can subsequently extend to others in certain circumstances. This appeal to particularism, however, does not resolve the question of foundation in Walzer's work. Walzer starts with the communal subject but this doesn't explain how this subject comes into being. Therefore, we must look at the articulation of Walzer's broader communitarian project in order to understand how community, which creates the possibility of both maximal and minimal morality, is founded.

Self-determination and membership

The prologue for Neil Gaiman's *The Doll's House* (2010) recounts the meta-narrative of community, the tale of how the story of community's origin is passed down the communal lineage. Gaiman describes the ritual iteration of a communal origin: a boy on the cusp of manhood is brought to the barren centre of the desert by a male relative to hear the tale of who his people really are.[3] The telling of the story is a performative and constitutive exercise: performative in respect of the pedagogical roles the participants play; and constitutive, because it is the telling of the story itself that completes the communal subject. A boy leaves to hear the tale but a man returns to the tribe: 'When he returns to the tribe he will truly be a man: he will have heard the tale. At night he will sleep in the young men's hut' (2010: 15). The man who returns is entrusted with the continuation of the narrative. The communal subject is duty bound to repeat the ritual later in his life, a circular motif that is the continuation of community itself. Gaiman's story recalls Jean-Luc Nancy's (1991) depiction of the mythical scene of communal foundation. Nancy argues that the mythical scene of community symbolises the desire to trace the lineage of community back to a singular starting point in which the retelling of the origin story is pivotal. This ideal of community is auto-affective: community is created and sustained by the self-repetition of the story of its origin. In other words, the auto-affective mythos proclaims that community gives birth, and re-birth, to itself through the articulation of a narrative: the tale of who *we* are, where *we* come from and what life means to *us*.

This image of community is fundamental to Walzer's ontology. Walzer maintains that self-determination is the primary condition necessary for communities to produce their own unique articulation of society – their maximal world. In this respect, Walzer is telling us that authentic communal life must be created by members of the community. The exclusion of alterity is necessary to maintain this foundation because the possibility of a *self*-determining community is underpinned by the assumption that there are others who are outside and not part of the self being determined. In Walzer's terms, in order for members of a community to build a particular maximal world they must be separated from strangers who do not share their maximal life. Walzer's justification of violence is intimately tied to the member/stranger dichotomy. Walzer argues that war is justified when a community's common life is threatened by nefarious border crossings of strangers. The crime of war is defined as the point at which a stranger threatens to cross the border and violently change a community's common life (2006a: 51–53).[4] For Walzer, intrusive strangers threaten to destroy the common life by illegitimately changing social meanings: 'Tyranny is always specific in character: a particular boundary crossing, a particular violation of social meaning' (1983: 28). Thus, the purity of the political community and the meanings it shares is threatened by what lies outside its borders. Self-determination, the core principle of Walzer's ontology, takes on a rather literal meaning: the self must be able to determine itself free from the coercion of others.

Before we begin to unpack the implications of Walzer's justification of war, it is perhaps necessary to briefly reiterate Walzer's maximal morality. As previously explained, maximalism describes morality shared between members of a political community. Walzer explains that shared values are the result of cultural memory, customs and shared social goods that coalesce into what he describes as a common life (1994: 8). As such, maximal morality derives from a collective historical process. For Walzer, community is the space in which maximal morality comes into being: 'the political community is probably the closest we can come to a world of common meanings. Language, history and culture come together to produce a collective conscience' (1983: 28). Although Walzer is keen to stress the commonality of maximal meanings, nevertheless, this commonality should not be mistaken for stasis. Social meanings are not simply agreed once and for all; they are fluid, always open to dispute and reformulation (1994: 27). This provides us with a clear impression of how Walzer envisages community to operate: a collective of culturally and historically related people deciding how they want to live together. Meanings are shared to the extent that members can understand and debate their common life. While members may disagree on the destination of this common life, their shared meanings, however, allow them to disagree while speaking the same vocabulary. This image is in contrast to the global community, which Walzer assures us has members but no history, culture or shared understandings (1983: 29–30). Running through Walzer's entire *oeuvre* is an understanding of the world in which humans exist within communities that possess distinct shared lives. Thus, Walzer is acknowledging that without a distinction between those inside a community (members) and those outside (strangers), common life and shared meanings would be impossible.

Pin-Fat (2010) astutely identifies that community represents something of a universal container in Walzer's argument, because without communities there would be no way for maximal or minimal morality to emerge. In this respect, the space of community represents another iteration of Walzer's particular/universal ontology. While Walzer stresses the particular character of the divergent communities that emerge within such spaces, the blanket potentiality for common life can only emerge within a universalised form of bounded space – the state.[5] Indeed, bordering principles define every level of Walzer's ontology, from his view of the international, to the domestic, to the subject. In short, Walzer argues that a bounded spatial plane, either physical or metaphysical, is a necessary condition for meaningful existence.

Maximal life is dependent on the existence of separated states. In turn, Walzer asserts that communities are forged through distribution, as 'we come together to share, divide and exchange' (1983: 3). Walzer expands on this image by arguing that communities are defined by a series of, what he calls, distributive spheres: for example, educational, economic, health and political spheres, and so on (1983). Walzer explains that understanding a particular set of distributions is tantamount to understanding a community's social character: 'Different goods in different companies of men and women for different reasons and in accordance with different procedures. And to get all this right, or to get it roughly right, is to map out the entire social world' (1983: 26). In this sense, maximal morality is tied to the ways in which a society collectively understands the meaning and distribution of its social goods. More specifically, Walzer maintains that goods have shared meanings because they are the result of socio-historical processes – and this is the reason that goods have different meanings in different societies (1983: 7). Distribution constitutes the underlying structure in Walzer's depiction of community; it is inseparable from morality because moral argument is simply an appeal to the common meanings that distribution creates (1983: 29). Once again, Walzer is restating the minimal/maximal idiom in a different context. Distribution is a minimal principle of all humanity, but the specific ways in which goods are distributed constitutes their maximal articulation; the universal container facilitates the production of particular content. In addition, Walzer proposes a dual defence of distributions from injustice. Internally, politics ensures that no single distributive sphere is dominant over others and war protects communal meanings from external destruction or alteration (1983: 16). These internal and external gatekeepers are crucial to Walzer's understanding of a just society: war protects the ability of a community to construct social meanings without external intrusion (self-determination); and politics ensures that social meanings are agreed within the appropriate distributive sphere, rather than enforced by a single dominant sphere.

We are beginning to grasp the auto-affective drive evident in Walzer's thinking. Authentic social meanings can only emerge from within a community of members and it is morally unacceptable for strangers to interfere in this process. In Walzer's terms, '[t]yranny is . . . to invade the sphere where another company of men and women properly rules' (1983: 18–19). For Walzer, membership defines who can *properly* engage in the construction of a community's common life: 'The theory of distributive justice begins, then, with an account of membership rights. It must vindicate at one and the same time the (limited) right of closure, without which there could be no communities at all, and the political inclusiveness of the existing communities' (1983: 63). Put in another way, membership founds social distributions – and therefore, the possibility of common life. Walzer, however, ambiguously situates the distribution of membership within the community itself: 'membership cannot be handed out by some *external* agency; its value depends upon *internal* decision' (1983: 29, my italics). In this sense, membership, which should rightly signify the origin and possibility of community, can only be distributed from within the bounds of a pre-existing community. Given that he is starting from a position in which people already exist within a community, it is unsurprising that, for Walzer, membership constitutes the foundational communal good: 'The primary good we distribute to one another is membership in some human community' (1983: 31). Walzer argues that admission and exclusion represent the core of communal independence and suggest the deepest meaning of self-determination (1983: 61–62). In other words, a self-determining community is a community that has power over membership – power over who is included in and excluded from the distribution process. In being presented with this central role of membership to Walzer's argument, it is important that we investigate the process of how someone becomes a member of a community. Walzer accomplishes this by re-iterating the argument that membership is a gift that is offered to the outside by the inside:

> . . . we who are members do the choosing, in accordance with our own understandings of what membership means in our community and what sort of community we want to have . . . we do not distribute it among ourselves; it is already ours. We give it out to strangers.
>
> (Walzer 1983: 32)

Again, Walzer describes membership as pre-existing: you are either born into membership or you are granted it by those who are already part of the community.[6] Before we begin to tackle the problem of the foundational distribution of membership, it is important to briefly discuss the ways in which strangers are invited to become part of an existing community (1983: 33).

Walzer begins his discussion by assuring us that those inside a community define what membership means and have ultimate authority over admissions policies (1983: 43). This position is clarified with the analogy that communities are like perfect clubs with full control over the selection process (1983: 40–41). In this respect, members are said to decide upon admissions in a free manner. However, Walzer almost immediately limits this idea by arguing that communities recognise a 'kinship principle', which gives membership priority to national and ethnic 'relatives' (1983: 41). He presents us with the image of freely distributed membership and almost instantly qualifies this free deliberation with the criterion of kinship. Walzer conceives the distribution of membership as a form of hospitality, a welcoming of particular strangers into the community. Yet the welcome is given precisely on the grounds that the stranger already shares a familial bond; we only admit others who are already like us. In Dan Bulley's questioning of European Union membership, he argues that such reasoning constitutes a nullification of hospitality, because it attempts to transform the other into the same before admission is granted (2009: 72). Bulley's statement is of great importance to the way in which Walzer foresees the distribution of membership. Walzer suggests that communities should be receptive to alterity, what he describes as a model of 'limited closure'. Nonetheless, he expresses this reciprocity in terms of openness to the same or the similar. Communities should not welcome strangers into their boundaries. They should invite relatives precisely because of pre-existing kinship. As such, Walzer presents us with ethical responsibility defined in terms of the self's recognition of itself in the other. Communities offer welcome to strangers who share a kinship. Ethical responsibility begins in the self and then expands to incorporate others who are like the self.

Nevertheless, Walzer does not necessarily view conventional requests for membership as a moral imperative. In fact, throughout *Spheres of Justice*, he implies a more utilitarian understanding, in which we invite strangers on the condition that they complement and benefit the community. Walzer does, however, acknowledge a moral dimension to appeals for membership when discussing refugees. He describes refugees as stateless people who endure an existence of 'infinite danger' and therefore require an ethical response (1983: 32). He asks, how should communities respond to refugees' pleas for sanctuary? Walzer's answer is an extension of the kinship principle. He argues that we are bound to help refugees 'if they are persecuted or oppressed because they are like us' – adding that we can share ideological kinship in addition to ethnic kinship (1983: 49). Once more, Walzer presents us with ethical responsibility based upon self-identification. The refugee is not simply accepted on the basis of his/her need or peril, but on what Walzer calls 'a sense of relatedness and mutuality' (1983: 50). Walzer's central argument is that the ability to control borders is essential to self-determination; thus, it is crucial that the outside we admit is already related and amenable to the inside. We exclude those who are not like us and do not share our communal values, even if they are in need of refuge. It is only through demonstrating ethnic or ideological kinship with existing members that strangers can be accepted. This, in turn, ensures that communal meanings remain internally coherent. The outside that Walzer desires to admit does not disrupt the inside because it is already part of it.

Declaration and the birth of community

The previous section highlighted the importance of membership to Walzer's understanding of self-determination. Membership determines who can negotiate shared meanings and the types of strangers who can join the community. At its base, membership is the foundation of a communal home that can house values, meanings and mutually related people. Derrida contends that the concept of home is a fundamental necessity for the possibility of ethics, because it defines who we are and how we can relate to others (2009: 16–17). The home is of similar importance to Walzer as it signifies the space in which members exist separate from, and are therefore able to relate to, strangers. Following his terms, the home constitutes the practical reality of existence: 'For citizenship entails what we might call "belongingness" – not merely the sense, but the practical reality, of being at home in (this part of) the social world' (Walzer 1983: 106). In Walzer's argument, the home's essence is located in members' ability to decide their own destiny without external intrusion, and alterity is conceived as always already outside this process rather than something implicated within. Nonetheless, Walzer's depiction of how communities are founded calls this ideal of self-determination into question.

Community is defined by the common life of members, the meanings they share and the distributions they negotiate. Yet the distribution of membership already presupposes the existence of a common life; the statement '*we* give it out to strangers' presupposes that there is an already existent commonality prior to the distribution of membership (Walzer 1983: 32, my italics). Walzer's ontology starts from a position in which people already exist within communities defined by shared values. Members are members because they share a common life and strangers are strangers because they do not share this particular social world. However, Walzer also maintains that a community's common life and shared values are the result of long historical, social and cultural processes driven by members. In other words, membership is necessary for a community to build a common life, but membership is also defined by a shared common life. In this way, membership and common life combine to create an auto-affective origin of community. In Walzer's terms, 'the common life is simultaneously the prerequisite of provision and one of its products' (1983: 65). We require a common life to distribute goods – and most importantly, the good of membership. Nevertheless, the common life is itself a product of distributions. In Pin-Fat's words, Walzer's account of communal origin 'presupposes the very thing it is supposed to account for' (2010: 90). Walzer's account of membership and common life presupposes that we are always already in possession of these things. His description of the foundation of community presents us with a metaphysical dead end, a non-origin at the origin of community. It becomes impossible for us to separate or distinguish between membership and common life. Both concepts are infused in an indeterminate account of how community begins and form a chain of supplementary origins. This is what Derrida means by the law of supplementary commencement: when we go looking for Walzer's origin of community, we are faced with a non-origin, a community that is already home to members who share a common life; a community already in motion.

To explain this argument in further detail, it is useful to provide a more illustrative example. In this respect, the US Declaration of Independence provides us with a clear articulation of communal self-proclamation. Walzer describes community in terms of a Rousseauian social contract: '. . . over a long period of time, shared experiences and cooperative activity of many different kinds shape a common life. "Contract" is the metaphor for a process of association and mutuality' (2006a: 54). For Walzer, the signing of the social contract indicates a willingness of a group of people to collectively decide how to distribute goods and build a common life (1983: 65). In short, the social contract symbolises the opening words in any maximal morality. The ethos of Walzer's depiction of social contract is evident within the spirit of the Declaration, which is above all a will to self-determination. John Shy reminds us that the Declaration 'was intended to foreclose serious negotiations which the British seemed ready to undertake' (1976: 11). Importantly, the Declaration also provides an instance of a community asserting the inauguration of itself in the name of itself. As such, the Declaration appears to conform to Walzer's ideal of community: a pre-existing mutuality committed to realising itself. The Declaration, nonetheless, proves to be a far more disjointed construct than Walzer's conception of community implies. Although the Declaration symbolises the annunciation of a self-determining subject, the authority through which self-determination is derived is divided. The opening paragraph implies that the authority to break from Britain derives from God and nature: '. . . to assume the powers of the earth, the separate and equal station to which the laws of nature and of nature's God entitle them' (*US Declaration of Independence* 1776: 1). This assertion of divine or natural authority stands in stark contrast to the appeal in the closing paragraph, which declares independence 'in the name, and by the authority of the good people of these colonies' (1776: 4). The concluding paragraph mirrors Walzer's ideal of auto-affective self-birth: communal foundation achieved through the commitment of a group of related people to attain self-determination. Nevertheless, the opening appeal to divine authority suggests that the new community lacked the absolute authority to inaugurate itself. The divided seal comes together in the proclamation of the truths of US community:

> *We* hold these truths to be self-evident, that all men are created equal, that they are endowed by their *Creator* with certain unalienable rights that among these are life, liberty and the pursuit of happiness.
> (*US Declaration of Independence* 1776: 1, italics mine)[7]

Within this sentence, the twin structures of divine and auto-affective authority come together: the commitment of the people to build a common life is underpinned by divinely given rights. Yet this combination creates an ambiguity surrounding the foundation of communal authority.

Hannah Arendt's (1963, 1970) reading of the Declaration emphasises the auto-affective aspects of the text: she regards the appeal to God as an unnecessary impurity in the founding of a new form of authority. Bonnie Honig explains that for Arendt, the Declaration symbolises the free coming together and public expression of a desire to build a community: 'The *We hold* is a promise and a declaration; it signals the existence of a singularly human capacity: that of world building' (1991: 101, original italics). Arendt's account is strikingly similar to Walzer's ideal of social contract: the Declaration is signed in the spirit of a commitment to build a community together. In Walzer's words, '[i]t never happened that a group of people called Americans came together to form a political society called America. The people are Americans only by virtue of having come together' (1996: 27). For Arendt and Walzer, the act of coming together signifies membership founded on the promise of the creation of a common life. Authority, in turn, is derived from the promise of community itself, and the commitment to build a common life becomes the foundation of this common life. However, for this account to make sense it relies upon the assumption that the people who came together to found the US already possessed a degree of shared understandings. In Walzer's terms, the commitment to build a common life presupposes that 'Americans' had a shared understanding of what their coming together meant, and that it signified the desire to build a maximal world. In other words, he presupposes that mutuality, a community of would-be 'Americans', existed prior to the coming together.

For Derrida, it is precisely the status of pre-existing mutuality that necessitates the supplementary appeal to divine authority. What is crucial to Derrida's reading of the Declaration is the status of the 'we' that performs the promise; that is, the status of the presupposed community:

> The 'we' . . . does *not* exist, *before* this declaration, not *as such*. If it gives birth to itself, as free independent subject, as possible signer, this can hold only in the act of the signature. The signature invents the signer.
>
> (Derrida 1986: 10, original italics)

By calling into question what Arendt and Walzer assume in the foundation of US community – that is, the pre-existence of the communal subject – Derrida highlights the ambiguity of the *we* who signs for the Declaration at the moment it pronounces its own birth. Importantly, Derrida argues that the authority to sign for independence is derived from the people who are declared independent in the act of signing. The act of signing is therefore simultaneously a performative and constitutive exercise: it is performative, because the promise of a commitment to the new community signifies the possibility of the *we*, who is capable of signing; and constitutive, because the signature retroactively produces the *we* who finally signs. It is only after the signing is completed that we can even claim to distinguish between the performative and constitutive aspects of the foundational act.[8] In this respect, the Declaration itself produces the US people who retroactively sign for their own independence. The Declaration is thus a retroactive justification of their authority: the signature produces the subject who provides the authority to sign for independence in the first instance. In Walzer's terms, we are faced with an ambiguous relationship: between the membership that creates the common life and the common life upon which membership is presupposed. In this way, the birth of community, which for Walzer is always in terms of a subject who can singularly sign for collective action, is fractured by the retroactivity of self-birth.

It is in the ambiguity of *ex post facto* foundation that the appeal to God becomes necessary.

Membership and alterity

Walzer conceives the foundation of community in the manner of auto-affective subjectivity: community comes from within itself and is founded by the promise of world building. In Walzer's argument, what is other than community is at best unnecessary and at worst threatening to community's very existence. Importantly, what is designated as other than community is pre-conditionally situated outside the community and members' common life. In the Declaration, the foundational promise of the US community is presented in the form of the commitment to be other than Britain. Similarly, Walzer defines members of the US community as 'voluntary immigrants' committed to escaping the old country (1996: 3, 28).[9] Accordingly, the US members are characterised by their voluntary escape from their former identities. The commitment to become 'other than . . .' becomes the conjoining feature of US solidarity. In Walzer's terms, the US experience is that of leaving a homeland and coming to this new place (1996: 17). The emergence of the US community, therefore, displays an interesting dynamic: members leave their former home because they do not properly feel at home, and find their new home by becoming other than what they once were. Walzer, in addition, contends that the emerging US community was unburdened by the cultural hegemony evident in more traditional societies, and this resulted in the formation of a community that was particularly receptive to difference (1996: 23–49). By leaving their old homes in order to find their proper home, members of the US community forged a particularly just society: 'it is one of the world's better societies: open, pluralist, and (relatively, again) egalitarian' (1996: 3). Yet it is precisely in the movement toward the new home that the distinction between inside and outside breaks down. Walzer's argument suggests that those inside their old home did not feel fully or sufficiently at home in what was their proper home. Despite being *de facto* members, they did not feel that their own maximal world represented their real maximal values. 'Americans' could only feel properly at home by building a new community; and their combined commitment to build this new home signified its foundation. In turn, it is only by virtue of expelling its former inside, the administrative connection to Europe, that the US community could be authoritatively announced. That is, the US became a subject that could singularly sign for its own destiny by rendering its colonial inside as a constitutive outside.

It is within this confusion and ambiguity between what is properly inside and outside US community that the divided seal of the Declaration begins to make sense. Walzer acknowledges the divided seal of US identity in terms of what he describes as *hyphenated being*: for example, 'Italian-American', 'African-American', etc. He argues that US identity is defined in terms of a lack of inwardness, a lack of historical, social and cultural commonality (1996: 26). Because of this lack of inwardness, the founding members had to look backwards, to their old countries, to find their maximal values. This creates a rather fractured image of US community. The US came into being in the form of a collective escape from the old countries; but the new community retained part of what it once was, because it lacked the culture and history necessary to forge a common life. Thus, that which the US rendered as its constitutive outside and forcibly expelled in the Revolutionary War, its colonial heritage, was simultaneously retained as a necessary component of the inside. The outside is retained in the new system as, what Derrida would term, a necessary parasite (1988: 90). The retention of former maximal identities was necessary because they made US community possible. Yet they disrupt the new community because they are intimately tied to the old regime. As such, the outside of the US is constitutive in two senses: it defines the system that the new community is moving away from and must expel from its inside; and the retention of the outside marks the very possibility of this movement. This signifies another iteration of the law of supplementary commencement. The US community is inaugurated as a self-determining subject only within the dynamics of a constitutive relationship with an alterity that is expelled from, and simultaneously retained in, the inside. In Walzer's system, the US retains its former cultures but expels its prior politics: 'If the manyness of America is cultural, its oneness is political' (1996: 29). In this way, Walzer's hyphenated identity functions by way of a cultural multiplicity (Italian, African, etc.) conjoined by a political singularity (American). Walzer asserts that US politics represent a break with the British system and this break is tantamount to the acceptance of the ideals of liberty, equality and republicanism (1996: 30). Hyphenated identity captures the twin principles of expulsion and retention and, for Walzer, constitutes a surplus in which US politics complements the pre-existing cultural plurality (1996: 45). In this way, Walzer attempts to domesticate the outside by the deployment of the hyphen. Rather than addressing the tension between the cultural multiplicity and political singularity, the hyphen creates a mythical separation of the terms which aims to fix the outside as a pure and simple addition: the cultural multiplicity is added to the political singularity.

However, the hyphen also helps us to locate the 'we' who authorises the Declaration of Independence. In Walzer's analysis, the US shares a common politics that facilitates the co-existence of multiple cultural identities. This common politics means that anyone can, in principle, become a member of the US community. In Walzer's words, it is '[p]recisely because the United States was no one's *national* home, its politics were universally accessible. All that was necessary in principle was ideological commitment' (1996: 35, original italics).[10] Recalling that mutual commitment to create a common life signifies Walzer's conception of communal foundation, the Declaration can now be re-read as a commitment forged on the basis of common politics and ideology. But this presents us with a foundational problem, because the common politics upon which the commitment is premised is announced in the Declaration. If the US community breaks politically with the British system, then it is precisely at the moment of the Declaration that this break occurs. Yet the pronouncement of the new political system is itself underwritten by an appeal to an existing ideological commitment of the people to the principles of life, liberty and equality. Read through Walzer, the Declaration presupposes the existence of that which it seeks to announce: a common politics underpinned by an ideological commitment to liberty, equality and republicanism. In this respect, Walzer is suggesting that the US community was forged by the coming together of people already committed to shared understandings of the principles of life, liberty and equality: people who shared a form of maximal morality.

It is in acknowledging this dimension of the Declaration that the necessity of the appeal to God and nature becomes apparent. In Walzer's argument, the commitment to build a community based upon shared understandings of life, liberty and equality symbolises the foundation of US maximalism. However, the shared ideology presupposed in this argument cannot exist prior to the foundation of the community because the ideological mutuality has yet to be forged through historical, social and cultural processes. It is for this reason that the signers of the Declaration underwrite their political ideal by appealing to natural and divine laws. Derrida conceives this appeal in terms of a place holder, because it sustains the authority of the Declaration until its task of creating the signer is completed:

> He [God] founds natural laws and thus the whole game which tends to present performative utterances *as* constative utterances.
>
> (Derrida 1986: 11, original italics)

The absence of a communal subject who can sign for the Declaration marks the absence of the common politics that Walzer's understanding of membership presupposes. The writers of the Declaration appeal to God to sign for the ideological values (life, liberty and equality) upon which their politics is founded, until such time as they can build a common life together. As such, US maximal morality is founded upon an appeal to the theological, which is in truth an appeal to absolute alterity. In Derrida's terms, the Declaration is 'a vibrant act of faith' (1986: 12). God signs in place of the community, as its place holder, until a common politics arises that can retroactively seal the contract. Self-determination is guaranteed by an appeal to absolute alterity.

Derrida contends that the foundation of community is implicated in the performative and constitutive structure of revolutionary violence.[11] He argues that revolutionary violence exists within a legal vacuum in which the old system is renounced and the new system has yet to be legally inaugurated:

> It is in the moment in which the foundation of *droit* remains suspended in the void or over the abyss, suspended by a pure performative act that would not have to answer to or before anyone. The suspended subject of this pure performative would no longer be before the law, or rather he would be before a law still undetermined, before a *droit* still nonexisting, a *droit* still ahead, still having to and yet to come.
>
> (Derrida 2002a: 270)[12]

The performative violence Derrida speaks of is tied to the performative Declaration of Independence. Violence creates the law and the subjects of this law who can retroactively sign for their own independence. The US War of Independence transformed the population from British subjects to citizens of the United States subject to US laws and political authority. More directly, Derrida argues that revolutionary violence creates a new system of law that retroactively justifies the violence that was instrumental in its own production (2002a: 269). As Shy affirms in the case of the Declaration, the Revolutionary War is justified in the US imagination because it created a free nation: 'Whatever was done or decided in 1775 or 1777 or 1781, the outcome justified it . . . The American nation was a success story from the beginning' (1976: 9). In other words, the violence that created the new political system, a common US politics, is retroactively justified in the name of what it created. Revolutionary violence produced the signature of the Declaration that retrospectively justified all that was done in its name, including the violent production of itself. Violence founds and signs for maximal morality through the constitutive exclusion of the new outside of US community: British political authority.

War and maximal morality

The previous section illustrates why violent exclusion is fundamental to the production of maximal morality. Walzer, however, is keen to emphasise a more peaceful image of community. He argues that a just society is one that lives life according to members' shared meanings: 'A given society is just if its substantive life is lived in a certain way – that is, in a way faithful to the shared understanding of its members' (1983: 313). The question, then, shifts to how we can ensure that the common life is authentically created and not tyrannically imposed. Walzer addresses this in a discussion on the relationship between maximal morality and interpretation. Interpretation is contrasted to what Walzer describes as the two alternative moral schemas of 'discovery' and 'invention'. Moral discovery is disregarded because it requires God to reveal the moral language to us: 'someone must climb the mountain, go to the desert, seek out the God-who-reveals, and bring back his word' (1987a: 4); while moral invention is disqualified because the inventor assumes the role of God: 'they create what God would have created if they were a God' (1987a: 12). Ultimately, Walzer asserts that we do not need discovery or invention, as we already have what they pretend to provide; interpretation allows us to debate actual existing morality (1987a: 21). Once again, Walzer argues on the basis of reality. The divine world of discovery and the mythical world of invention are unnecessary in the face of interpretation, our experience of actually existing real world morality. Nonetheless, an interpretative real is also a contested real. How then do we know if maximal morality genuinely represents a community's shared understandings? As stated previously, Walzer claims that the sphere of politics can guarantee the fidelity of maximal morality and shared meanings. More specifically, Walzer presents politics as a means to police the boundaries between the various spheres of distribution that constitute a community's common life:

> It is used to defend the boundaries of all the distributive spheres, including its own, and to enforce common understandings of what goods are and what they are for . . . political power is always dominant – at the boundaries but not within them.
>
> (Walzer 1983: 15)

In this way, Walzer places the sphere of politics at the centre of communal life. Politics polices the boundaries of maximal morality by ensuring that life is lived according to members' shared values. Yet Walzer also acknowledges that political power often oversteps its remit by breaking into the distributive spheres and changing social meanings, rather than defending them (1983: 282). Walzer's view coincides with the Derridean (2002a) understanding of police power. Derrida maintains that even though policing is designed to preserve the law, it always risks remaking the law through enforcement. In this sense, although politics is necessary to defend maximal morality, this preserving power risks violating shared meanings. Because of the unstable nature of police power, Walzer posits democracy as a supplement to politics. He argues that 'the only thing that can justify undemocratic forms of government is an undifferentiated conception of social goods' (1983: 303). Put differently, democratic politics is essential to preserve the integrity of distributive spheres. Definitively, Walzer argues that democracy is an essential atom of any decent society: 'I want to argue that a decent society requires not only individual rights but also group solidarities and the pluralist and democratic politics that make groups possible' (1996: 122). Walzer's understanding of justice is, in this way, intimately related to democratic politics.

Walzer characterises democratic politics as a public forum for debate:

> Democracy puts a premium on speech, persuasion, rhetorical skill. Ideally, the citizen who makes the most persuasive argument – that is the argument that actually persuades the largest number of citizens – gets his way . . . All other citizens must talk, too, or at least have a chance to talk . . . Equally important is what we might call the rule of reasons. Citizens come into the forum with nothing but their arguments. All non-political goods have to be deposited outside: weapons and wallets, titles and degrees.
>
> (Walzer 1983: 304)[13]

In this respect, Walzer posits an implicitly Westernised model of democracy as the mechanism through which politics operates in just societies. Politics, however, is now in danger of being supplanted by democracy. Walzer wants to ground community on the ideal of politics, the commitment to living together according to shared understandings. Yet, in order to create a just and decent society, politics must take a very particular form. Justice can only be derived through Western-styled democratic politics. In fact, Walzer acknowledges the particularist nature of this ideal of politics when critiquing the Habermasian democratic speech theory:

> For the minimal morality prescribed by these theories is simply abstracted from, and not very far from, contemporary democratic culture. If no such culture existed, this particular version of minimal morality would not even be plausible to us . . . very much like an oak tree that, endowed with speech and encouraged to speak freely, solemnly declares the acorn to be the seed and source of the entire forest.
>
> (Walzer 1994: 13)[14]

Walzer's critique is interesting because it implies that Western democracy cannot be viewed as a minimal condition for authentic maximal life. In acknowledging the particular nature of democracy, Walzer contends that authentic communities do not necessarily need to be just. Rather, to embody a maximal existence a community needs to be legitimate. The distinction between just and legitimate is important to Walzer's picture of the world. While an unjust society is cause for moral criticism – and even strangers are entitled to levy such criticism (1983: 314) – the absence of democracy alone is not sufficient enough to doubt the existence of community and common life. In short, the absence of democracy does not challenge a community's minimal right to determine its own maximal world. Therefore, it is the category of legitimacy that is important in regard to a community's claim to self-determination.

Maximal justice necessitates open public debate. Minimal legitimacy, however, is derived exclusively from a people's capacity for collective violence. Walzer grounds a community's legitimacy on what he refers to as the self-help test: the ability of a government 'to help itself against internal enemies' (2006a: 99). Walzer's depiction of the self-help test is relatively straightforward: if a government does not represent the true values of a community, the people will seek to overthrow it – and both resistance to the government and the punishment for this resistance are legitimate.[15] Ultimately, the side that can gather the most support for their cause represents the genuine maximal values of a community.[16] Walzer, however, does not view violence as a lamentable consequence of competing claims to self-determination. Instead, violence marks the presence of genuine politics – because if the issues are significant, violence is always a risk, and to remove this risk would reduce the political process to a charade (1996: 94). Walzer valorises the willingness to resort to collective violence in the defence of maximal meanings as a sign of authentic community.[17] While a system of democratic politics emphasising public discourse and the rule of reason is desirable, all communal legitimacy requires is a *coup de force*. The community, which above all exists to protect the common life from the external menace of war, is legitimised by that which it seeks to keep outside its borders. In Walzer's image of community, collective violence creates the maximal world and subsequently justifies itself solely in defence of the common life it has created.

To conclude our discussion on community, we will examine Walzer's conception of tyranny, the structure of an unjust society. Walzer describes tyranny as 'a continual grabbing of things that do not come naturally, an unrelenting struggle to rule outside own company' (1983: 315). Rather more succinctly, he clarifies that tyranny is simply 'the exercise of power outside its sphere' (1983: 59). Both of these descriptions rely on an understanding of life already anchored in the maximal world, a world divided into members and strangers. However, the violent inauguration of common life teaches us that the creation of the maximal world, its spheres and distributions, is grounded upon a performative and constitutive violence that operates outside all conventional structures. This founding violence is the most tyrannical, as it strives to rule outside all existing *droit*; yet it is simultaneously the least tyrannical, because it does not recognise any *droit* aside from that it creates (Derrida 2002a: 274). In other words, the violence that creates and legitimises a community and its common life is a form of tyranny that retroactively justifies itself through the system of law it creates. Walzer argues that the coercive transformation of a way of life is the death of the community (2005: 49), and that shared understandings cannot be the result of radical coercion (1994: 27). Yet the foundation of maximal morality is already radically coercive. In the example presented above, the creation of a new common life, of new laws and norms through the US Declaration of Independence and the Revolutionary War, was both radical and coercive: it pitted members against each other and the new political authority was cemented through militia violence (Shy 1976). What Walzer would ordinarily term tyrannical violence – that is, the creation of a common politics through means of violent exclusion – sets the maximal world, and therefore his entire ontology, in motion. Walzer declares that politics is the cornerstone of maximal morality, protecting common meanings and defending the boundaries of common life without violating them (1983: 15). Nevertheless, the possibility of this boundary defence is predicated upon the violent demarcation of boundaries between members and strangers – the violent production of community.

Walzer legitimises the violent and coercive creation of a common life when it derives from self-determination. However, the present analysis has illustrated how foundational acts of violence are in themselves productive of what constitutes inside and outside for a community. In the US example, prior to the Declaration there was no common politics distinct from the British system. In Walzer's own terms, there was no maximal bond or even people of 'the United States'. The common politics – the community – announced in the Declaration was created through a revolutionary war that coercively transformed the whole population's way of life. The revolutionary militias instituted a new system of laws and norms through coercive threat and use of force. The inside-of-community, as such, is itself a product of coercive violence that stands prior to the institution of maximal morality. Walzer ties communal legitimacy to the ability of governments to defeat their internal enemies. Yet the very possibility of the distinction between inside and outside is underpinned by a constitutive violence that refuses to recognise any system of law aside from what it creates. In this sense, what threatens community – the violent transformation of the common life – threatens community from the inside. In the foundation of US community we see the breakdown of the distinction between inside and outside. Europe, which resides at the heart of US cultural heritage, was constituted as an external stranger through revolutionary violence. It is only by rendering its inside as an outside that community can found the member/stranger distinction, maximal morality and the entire ontological game. War tyrannically founds community, its boundaries, distributions and common life.

Temporal revelation and being

Walzer grounds communal legitimacy in the war and revolutionary violence that produces self-determining subjects capable of constructing maximal moralities. In Walzer's model, the self becomes itself in relation to what it is not. The subject emerges by conceiving alterity as a constitutive outside. By showing the impossibility of the maximal foundation of morality we call into question the possibility of universal minimalism. Yet the critique of maximalism does not necessarily discredit its functionality as the means through which the minimal rules of war are produced. In other words, the violent and unjust foundation of the maximal world does not preclude the possibility of communities negotiating minimal rules of war. Importantly, the critique of community does not tell us how or why minimal morality fails to function in Walzer's posited ontological system.

Walzer unveils minimalism by presenting us with the image of protesters in Prague during the Velvet Revolution of 1989 carrying signs demanding 'truth' and 'justice':

> I knew immediately what the signs meant – and so did everyone else who saw the same picture. Not only that: I also recognised and acknowledged the values that the marchers were defending – and so did (almost) everyone else.
>
> (Walzer 1994: 1)

Walzer describes minimalism as a form of temporal revelation: minimal values are recognised within specific politically charged contexts. In Walzer's words, moral language reveals itself thinly on special occasions and we know minimal morality when we see it (1994: 4).[18] However, this is not an unproblematic argument because Walzer contends that minimal morality can never be actually expressed minimally; it can only be stated maximally:

> Minimalism when it is expressed as Minimal Morality will be forced into the idiom and orientation of one of the maximal moralities. There is no neutral (unexpressive) moral language.
>
> (Walzer 1994: 9)

Walzer's description of minimalism is, in one sense, a restatement of his belief that minimalism comes from maximalism. Yet in this case, the emergence of minimalism does not revolve around the negotiation of common values between communities, as implied in Walzer's description of the War Convention. Instead, temporal revelation constitutes an intimate, passive and spontaneous recognition of minimal values within maximal language. The people viewing the Prague protests did not need to know the cultural meanings implied in the signs because they recognised the underlying essence embedded in the gesture. In other words, although the signs had a distinctive cultural resonance understood within Czechoslovakia, they also contained a universal undertone that was recognised across all communities. In a similar manner, Walzer's claim that US politics is accessible to all peoples suggests that there is something about the ideals of life and liberty that renders them intimately comprehensible to every human being. The most striking example of this form of minimal revelation is Walzer's depiction of acts 'that shock the moral conscience of mankind' (2006a: 107) – atrocities that are so heinous that people are moved to respond, despite there being no direct threat to their own community. Walzer posits genocide and mass enslavement as examples of acts that shock our moral consciences. This depiction represents a shift in Walzer's conception of morality. He is no longer talking about the maximal morality that embodies a community's collective conscience or the minimal rules negotiated between communities. Instead, Walzer is now directly discussing the possibility of a universal conscience attentive to particular minimal values in specific instances. This image is unsurprising, given Walzer's belief that the values underlying minimalism are attached to our sense of what it means to be human (2006a: 54). As such, Walzer is suggesting that all humans share the capacity to be shocked by specific acts because we collectively recognise that certain universal rights are being violated en masse. Nevertheless, our recognition of minimalism does not represent a full-bodied universal morality. Rather, Walzer presents us with a universal morality that is recognised within a particular maximalist expression and subsequently interpreted through the individual's own maximalist vocabulary. In Walzer's (1994) terms, we may briefly join the minimal parade but we soon find ourselves back in our own maximalist one. In this respect, Walzer describes a form of universal morality inherent to the essence of mankind that, although silent and essentially unsayable, can be innately recognised in a myriad of maximal languages. Minimalism is therefore a spark embedded and recognised in all moral languages.

For Walzer, what is universal is not a minimal language but our ability to recognise minimal rights, and this recognition is possible because we are all human beings. Therefore, to understand the revelation of minimal morality we must discuss Walzer's conception of subjectivity: the mechanism through which humans recognise and interpret the moral world. Walzer describes the subject as an *ordered self* and contrasts this to religious conceptions of self that suggest God has placed a singular conscience in all of humanity (1987b: 33–43).[19] He describes this ordered self as a complex maximalist whole, internally divided in interests but not utterly fragmented (1994: 85, 96). Walzer depicts the subject as a thickly populated circle with a core 'I' surrounded by its self-critics. This 'I' is characterised as a newly elected president, capable of summoning advisors, forming a cabinet and manoeuvring between its constituent parts (1994: 98–100). Although Walzer is keen to stress the maximalist character of being, Pin-Fat reminds us that the structure of Walzer's being is universal (2010: 97). Walzer assumes that all human beings are like this, and that every person is comprised of a president and his or her circle of self-critics. Walzer wants to assure us that difference defines the heart of subjectivity, that humans are particular. Yet how this difference is structured is the same in every human – the structure is universal. In other words, Walzer's self may be maximally constituted but its structure is minimally distributed. In his depiction of subjectivity, Walzer presents us with another universal container. The structure of Walzer's subject is, as Pin-Fat argues, socio-historically pre-existent; it 'is not dependent on time and place though its *shape* may be' (2010: 97). Walzer's universal structure mirrors his image of community. The subject is maximal because it is shaped by its internal critics (divided interests and specific socio-cultural contexts). Its organising principle, however, is the same in all cases. The subject's structure is dictated, a priori, in a minimal way.

The subject is constituted by different presidents and different critics, but it is always already organised in this way. In Walzer's argument the maximally divided self is contingent upon a universal, immemorial internal structure. In fact, Walzer's ordered self strikingly resembles his image of democracy: a group of particular interests bounded within a secure space that are all afforded an equal opportunity to convince the community (in this case the sovereign 'I') that their interests should be adhered to. In this sense, Walzer's universal conception of subjectivity is endowed with a thoroughly maximalist character when judged by his own standards. Importantly, we must assume that this structure allows us to recognise minimalism because it signifies the common element inherent to mankind. The universal structure of the subject is the thread that links divergent social groupings together: people are similar because their internal structure is identical. It is through the democratically ordered self that the subject can reveal the minimal essence embedded in maximal communication.

Différance and secular theology

Walzer's overarching account of the minimal/maximal dichotomy brings us toward the Derridean concept of *différance* and the logic of the supplement. *Différance* is a play on the French word *différer* and its dual meaning, to differ/defer. Derrida argues that *différance* constitutes both a differing between meanings and a deferral of ultimate meaning; the delay inherent in signification and the difference that founds oppositional concepts. Derrida asserts that self-present meaning is the ideal of Western metaphysics; however, it proves impossible because *différance* inhabits the very core of what appears to be immediate and present (1981a: ix). He contends that in language, the sign, which is a representation of the thing, stands in place of the thing to preserve the thing's presence; but in doing so, it heralds the disappearance of the thing's natural presence: 'that what opens up meaning and language is writing as the disappearance of natural presence' (Derrida 1997: 159). In Walzer's model, for instance, we can only recognise minimal morality through maximal morality; yet the second we try to interpret minimal morality, it is already transformed into another maximalism; as soon as have we joined with the minimal parade, we already find ourselves back in our own maximal one. If universal values exist within Walzer's ontology, they are entirely unrepresentable and incommunicable in a minimal way. This has important implications because the rules of war outlined in the War Convention are professed to derive from these unrepresentable minimal values.

Derrida insists that every search for an origin, like our search for the origins of maximalism and minimalism, will find a non-origin. Invariably, what we will discover is not a singular starting point but a chain of supplements with meaning already contested at its roots (1997: 247). For Derrida, the supplement symbolises the relationship between the self and alterity that is productive of the distinction between these concepts. For example, the supplement is evident in Walzer's conception of community within the ambiguous relationship between inside and outside that founds the member/stranger distinction and the possibility of maximal morality. We never find the definitive origin of community. Instead, we find a process already in motion. In this respect, Walzer's thick and thin worlds are supplements for each other. In the first instance we are told that minimalism follows from maximalism. However, as highlighted in our discussion on the origin of maximalism, the only way a maximal morality is possible is through the detour of the supplement. Maximalism presupposes a universalised understanding community, described as a self-determining subject. Yet this universal structure is underpinned by a constitutive relationship between members and strangers that calls the possibility of self-determination into question. In the second instance, when we looked at the process of minimal revelation, we found that maximal expression is necessary for the articulation of minimalism – and that Walzer's universal depiction of the subject is premised upon democratic politics. Although these supplements threaten to usurp each other, they remain necessary for Walzer's ontology to function: 'a terrifying menace, the supplement is also the first and surest protection against this very menace. This is why it cannot be given up' (Derrida 1997: 154). In the absence of the law of supplementary commencement, without the constitutive interplay between self-determination (maximalism) and alterity (minimalism), Walzer can neither found community nor universal morality.

Not only does Walzer's conception of universalism display logical inconsistencies, in important respects it also has the characteristics of a theological model. In Walzer's description of minimal morality, minimal values are unveiled to us as a spark of recognition within maximal expression; and minimalism itself is eternally silent without the possibility of language or expression. In other words, the self-presence necessary to transform minimal values into clearly articulated moral laws remains deferred. Minimalism, in this sense, represents a secret revelation that takes place inside the subject that can never be outwardly expressed in its authentic form. The subject recognises minimal morality on the inside without the possibility of fully expressing it in a minimal way. Derrida discusses this theme in response to Søren Kierkegaard's image of subjectivity, describing God as the invisible interiority of the subject (2008: 108). This structure is of seminal importance to Walzer's conception of subjectivity. We never see the subject's internal mechanics, the presidential 'I', how it calls its cabinet together or how it recognises minimal morality in maximal language. As Walzer acknowledges, the production and reproduction of subjectivity is a great mystery (1987b: 43). Walzer attempts to refute the mystical, internal revelation of minimalism by arguing that the recognition of minimalism is a form of translation: we translate minimal values expressed in one maximal morality into another. This argument, however, cannot be sustained. Whereas translation requires a competent understanding of both languages, minimalism does not. In fact, the identification of minimal morality requires something that exists outside language to enable translation. It requires the recognition of something common to both languages but inexpressible in any language. It is only by assuming that the minimal value is recognised by everyone, members and strangers, that maximal translation becomes possible. As such, Walzer's argument is predicated upon faith in our ability to recognise unavowable minimal values, internally and in secret. Walzer's faith that minimalism is authentically recognised in maximal expression exceeds all onto-logical proof. Without this faith, minimalism could not possibly exist. In a similar manner to the Declaration of Independence and the foundation of US community, faith acts as a place holder that conjoins the transition of minimal meaning from one maximalism to another. Faith in minimal morality provides the link between disconnected maximal moralities and the possibility of universal rules of war.

This takes us to the crux of Walzer's secular theology. Walzer requires mini-malism in order to defend the type of universalism necessary for his rules of war. However, minimalism, as described by Walzer, is possible only through a move-ment of faith: faith that minimal values can be recognised within maximal expres-sion. Walzer weaves an onto-theological narrative that installs the War Convention as a universalised moral code written in absolutist terms, and he grounds this code upon the presumed existence of minimal morality. Yet this discourse can-not appear in the minimal dialect that Walzer requires and is thus a groundless foundation. Minimal morality is never present in any discourse on war or in any inter-communal argument. Walzer recounts the myth that minimal morality can be recognised within maximal moralities. Nonetheless, we never see or hear mini-malism, which ontologically exists nowhere. What we see, in Walzer's writings, is a maximalist War Convention that portends an inexpressible universal essence. In this sense, Walzer's minimal morality shares the characteristics of *différance*: its meaning is constantly differing across maximal moralities with its authentic meaning perpetually deferred. Therefore, it is only through a movement of faith that Walzer can claim that minimal morality is authentically represented in the mediating language of just war. Walzer's ontology requires us to have faith that minimal morality exists and faith that it is authentically expressed in the mediat-ing language of the War Convention.

Conclusion

Walzer's conception of wartime morality endeavours to provide a system of rules that are detached from the theological heritage of the just war tradition. He attempts to construct a viable rights-based universal morality that complements his overarching communitarian ideology. To this end, he builds the foundation of universalism upon the plateau of particularism: human rights that derive from communitarian ontology. The rules of war, Walzer's War Convention, sig-nify the practical articulation of this model of universalism. War, as a primary inter-communal engagement, requires the codification of universal laws to ensure violence is conducted within a moral remit. Walzer's ontology, nonetheless, can-not function in the way he professes it to: community and self-determination pre-suppose universal structures, and minimalism is silent in the absence of maximal articulation. Ultimately, this tells us that the world is not built upon self-determi-nation and particularism. Instead, Walzer's morality is founded upon a chain of supplements: the interplay between minimal and maximal, universal and particu-lar, and self and other, which are all necessary to co-found Walzer's ontologi-cal system. Yet it is precisely these relationships that preclude the possibility of

Walzer positing a universal real world morality. The War Convention is universal only to the extent that Walzer assures us that it preserves the essence of an inarticulate universalism. In this sense, universalism is, in Walzer's argument, embroiled in a movement of onto-theological faith: faith that what can never be present is, nonetheless, unwaveringly represented in the War Convention. Walzer's system of morality reassembles into a secular theology that pronounces a universal moral code through a movement of faith.

Walzer justifies war on the basis of a defence of communal rights. In turn, the defence of rights is underpinned by the protection of life and liberty: it is wrong to rob a person of life or to enslave them. However, these are the very prohibitions that are placed at risk during wartime: lives are placed on the line and freedom is called into question. In fact, Walzer argues that the fundamental crime of war is that it forces men and women to risk their lives in defence of their rights (2006a: 51–52). Therefore, absolute rights are not absolutely inviolable – in certain circumstances rights need to be risked in defence of community. Walzer thus presents us with a system of morality that originates in the community and satisfies itself in the defence of the community. In other words, our primary ethical responsibility is to defend the community, even if this necessitates the sacrifice of minimal rights: a just war is a war that preserves or produces an authentically self-determining communal subject. The purpose of the next chapter is to illustrate how this ideal of justice was implicated in the Bush Administration's justifications of the Iraq War and their broader post-war programme. In viewing justice in terms of the production of a cohesive socio-political unity, the US set events in motion that radically transformed Iraqi society and fundamentally changed the way in which Iraqis engaged with each other.

Notes

1 'Secular theology' should also be understood in terms of a supplement. It indicates a model that is neither properly secular nor theological, yet borrows from both as a means to co-found Walzer's ontology.

2 In this respect, the concept of community is integral to the possibility of nation states in Walzer's argument. Walzer explains that a nation state must already contain a community within it (1983: 44).

3 Gaiman stresses that the mythical scene is always a relationship between males. Women, Gaiman assures us, have their own stories, which tell a different tale.

4 Even when discussing humanitarian intervention, Walzer places the onus on preventing tyrannical rulers from destroying the community. Walzer argues that intervention can only be justified if a tyrannical regime is preventing the emergence of an authentic community; and once the tyranny has been averted, strangers must leave liberated members to work out the substantive content of their own community in the spirit of self-determination (2006a: 86–108).

5 Walzer argues that even though meanings are probably shared to a greater extent in smaller, familial groups, the state is the necessary space for community because it is the smallest possible formation that can protect common meanings from the intervention of strangers: 'To tear down the walls of the state is not, as Sidgwick worriedly suggested, to create a world without walls, but rather to create a thousand petty fortresses' (1983: 39).

6 Walzer argues that any attempt to substantially regulate membership of those born into a community would require excessive coercion (1983: 35).

7 Interestingly, these truths are intimately related to the absolute values of Walzer's minimal morality.

8 This theme will be revisited in the third chapter through a discussion on the foundation of conscious subjectivity.

9 Walzer excludes native peoples and those forcibly brought to the country as slaves.

10 This implies that US membership is determined solely on the basis of ideological kinship.

11 This type of violence was visibly evident in the formation of the US through the Revolutionary War against Britain.

12 Anidjar explains that the French word *droit* is notoriously difficult to translate into English. The word carries the sense of 'law' and 'code of law', and the sense of 'right' (as in 'the philosophy of right', but also the 'right to strike' or 'human rights'). It is distinguished from *loi* which signifies 'law' in the singular (Derrida 2002a: 230).

13 Note the masculine undertones in Walzer's conception of citizenship.

14 This quote is interesting as it disrupts Walzer's conception of hyphenated identity. Walzer's ideal of US identity rests upon the separation between the cultural multiplicity and political singularity. However, Walzer is now acknowledging that politics is itself cultural and historically produced, rendering the hyphen ambiguous at best.

15 This is provided that neither side engages in excessive violence or coercion that would constitute acts of genocide or enslavement (Walzer 2006a: 107).

16 See Walzer, *Just and Unjust Wars*, chapters 6 and 11.

17 In addition, it must be noted that Walzer does not completely bar strangers from engagement in the political process of a community of which they are not members. Walzer maintains that borders are not designed to keep out ideas; strangers have a limited right to speak and present their ideas to other communities. Foreign ideas, nevertheless, must be adapted by members to fit their cultural understandings. For example, Walzer argues that if he were to speak to a Chinese audience about democracy he would have to do so through the medium of US maximalism. However, if the Chinese found these ideas appealing they would translate his ideas in a way that made them amenable to Chinese values, culture and customs (1994: 58–61). As such, strangers are allowed to present their ideas and try to convince members of their merits, but they are denied the possibility of institutionalising their ideas through violence. In other words, membership is intertwined with the possibility of engaging in violence as a means to institute ideals.

18 Walzer describes these special occasions as personal or social crises, or political confrontations (1994: 3).

19 It is important to note that Walzer fails to provide any discussion on how subjectivity itself is formed. The ordered self is already a self-conscious, internally formed, subject. In a similar way to his depiction of a world in which people are always already members of a political community, subjects are presupposed to be composed as *ordered selves*.

2   Violence, ethics and the invasion of Iraq

A brief history of Iraq

The last chapter provided a critical analysis of Walzer's conception of ethics and politics. I argued that Walzer's understanding of justice and morality is guided by an image of the world in which the protection of collective self-determination is tantamount to the preservation of genuine politics and the fulfilment of ethical responsibility. Walzer conceives war as justifiable, in this respect, when it is designed to protect, preserve or restore self-determination to a pre-existing community of members. In order to contextualise Walzer's arguments and the consequences of his conception of ethical responsibility, this book will regularly turn to historical examples from the 2003 Iraq War and subsequent occupation. The purpose of this chapter, therefore, is to provide a brief overview of the 2003 conflict with the aim of illustrating how the US plan for Iraq is linked to Walzer's understanding of ethics. The goal of this chapter is not to recount *the* definitive truth of Iraq; rather, it will tell a particular story about Iraq that informs the philosophical and political discussion throughout this book. In this sense, this chapter draws together the problematic relationship between contemporary understandings of ethics and violent actions as a means to illustrate the real world implications of rights-based systems of morality.

The story of the 2003 invasion is in many ways inseparable from the contemporary history of Iraq and its seemingly unrelenting sequence of conflicts. Iraq's modern history, like much of the Middle East, is steeped in colonialism. In the aftermath of the First World War, Britain was awarded the mandate for control over Iraq. Iraqis, unsurprisingly, were unimpressed by the enforced arrangement and 1920 witnessed a brief unification of Shi'a, Sunni and Kurdish Iraqis during the 'Great Iraqi Revolution'. The revolution, however, was quickly quashed with Britain establishing a formal system of rule in 1921. While British troops initially claimed to come as liberators and not colonial conquerors,[1] the parameters set in 1921 largely remained unchanged until a 1958 *coup d'état.* Chief among the British system's legacies was the ascension of the Sunni minority to the role of ruling political class. Gregory (2004) explains that the British viewed the Sunnis as more amenable to control and actively funnelled political power to the Sunnis to the detriment of the Shi'a and Kurdish populations. Although this move was endemic of colonial divide and conquer techniques, its consequences have shaped and reshaped Iraq for almost a century.

Colonial rule played a major role in the cementation of Iraq's ethno-sectarian divide. Nevertheless, the pre-invasion politics of Iraq was more directly informed by the 1963 'Ramadan Revolution' and the Ba'ath party's seizure of political power. Adeed Dawisha (2009) explains that Ba'athism arose within the wider context of Middle Eastern decolonisation. The Ba'ath party claimed to offer the Iraqi people a more democratic form of secularised politics coupled with social-ist economic programmes. Ba'athist rule coincided with an economic and social renascence in Iraq. Fuelled by oil wealth, the Ba'ath regime invested heavily in infrastructure, education and the public sector, resulting in the emergence of Iraq as one of the most modern and economically developed nations in the region during the 1970s. Iraq's fortunes, however, began to turn when Saddam Hussein came to power. In 1976 Hussein ascended to the role of General of the Iraqi Armed Forces. Through his control of the military, Hussein soon gained overarching con-trol of Iraq's political institutions, establishing formal autocratic rule in 1979. In combination with repressive domestic policies, Hussein soon embarked upon an aggressive foreign policy stance, and the 1980 war with Iran marked the begin-ning of the end of Iraq's modern social and economic renaissance. Iraq's war with Iran is often articulated in terms of a continuation of the historical conflict between Arabs and Persians. Yet, the conflict was steeped in strategic and politi-cal dimensions. Dilip Hiro, for instance, contends that Iraq's main motivation was to regain land surrendered to Iran in a 1975 treaty (1991: 2). Iran was in the midst of deep domestic upheaval, and Hussein believed that Iraq could deal a crushing strike if they acted quickly. In addition, the 1979 Iranian revolution resulted in a theocratic Shi'a polity, and Hussein feared that the Shi'a majority in Iraq would attempt a similar coup. The war, in this respect, constituted an attempt by Hussein to protect his rule and expand Iraq's regional influence. However, it did not go according to plan. Iraq vastly overestimated Iran's perceived military weaknesses and the conflict descended into an eight-year bloody stalemate, absorbing billions of Iraqi dinars and killing hundreds of thousands of Iraqi soldiers. As Hussein diverted more and more resources to fuel the war machine, Iraq's social infra-structure stagnated. By the time a ceasefire was declared in 1988, the economic progress of the 1970s had been reversed, and Iraq emerged as a nation encum-bered by debt. It is in the context of this economic stagnation that Hussein turned his attention to Kuwait, a neighbouring country with vast oil reserves of its own.

While Hussein publicly declared that the 1990 invasion of Kuwait was an attempt to repatriate a former Iraqi principality, the economic toll of the war with Iran provided the more pressing impetus. The war had plunged Iraq into massive national debt, and Kuwait's ample oil supplies offered an immediate means to resolve this crisis. The story of the ill-fated invasion is well versed in Western nar-rative: an evil dictator invaded a small nation and an international coalition, led by the US, drove Hussein's forces back, saving the Kuwaiti people from tyranny. The political picture is undoubtedly more complex, but the net result remains the same: Iraq suffered a crushing military defeat and Iraqis were subjected to a prolonged international sanctions regime. Sanctions, in turn, further decimated Iraq and its people, reducing the nation to one of the most economically depressed and socially deprived in the world. Under sanctions, the Iraqi dinar dropped in value from one dinar equalling \$3.20 in 1990 to one dollar equalling 2,550 dinars by 1995; gross national product fell by 50 per cent during the first year of sanctions; and by 2000, Iraq was the third poorest country in the world. In 2000, senior United Nations (UN) official Rao Singh reported that some 500,000 children had died under the sanctions regime and in the same year, 25 per cent of all Iraqi children had dropped out of school for economic reasons and two million Iraqis, primarily from wealthier backgrounds, emigrated (Dawisha 2009: 123–128). Tellingly, by the end of the 1990s, 60 per cent of the Iraqi population were completely dependent on Oil-For-Food rations for daily survival (Napoleoni 2005: 143).

In 2003, US troops entered a country that was struggling for daily survival and almost entirely dependent on central government for jobs and daily sustenance. The US political mission, therefore, aimed at building the foundations necessary for Iraq to re-emerge as a modern, peaceful democracy in a region beset by violence. The purpose of this chapter is to illustrate how US political objectives radically altered Iraqi society. The chapter begins by outlining the relationship between Walzer's thought and the type of society the US invasion hoped to create. It then proceeds to unpack how the socio-political environment spiralled out of US control, creating a new Iraq distinct from Ba'athist rule but radically different from US pre-war imaginings. Although this chapter discusses post-invasion Iraq in terms of Shi'a, Sunni and Kurdish populations, it does not seek to represent these groupings as pre-given historical formations. Rather, it will demonstrate why the current articulations of Iraq's various ethno-sectarian identities are in part a product of the invasion and occupation. Iraq's current ethno-sectarian make-up, in this respect, is partly the result of the invasion and the ethico-political decisions made by the Bush Administration and the US Military.

The pre-war imagining of justice in Iraq

The Iraq War is often characterised as an exemplar of neoconservative foreign policy in action (see Schmidt and Williams 2008).[2] Indeed, Iraq had been firmly on the neoconservative foreign policy agenda throughout the 1990s (Hirst 2013). Advocates of intervention argued that Iraq had either procured or sought to procure weapons of mass destruction (WMDs) and that the Ba'ath regime posed a threat to security and stability in the Middle East. For example, a letter directed to then US President Bill Clinton by the Project for the New American Century (a prominent neoconservative group) argued for unilateral intervention on the grounds that '[t]he only acceptable strategy is one that eliminates the possibility that Iraq will be able to use or threaten to use weapons of mass destruction'.[3] Although neoconservatives were influential within the Bush Administration, the 2003 invasion of Iraq is inseparable from the September 11 attacks in 2001. These attacks confirmed to neoconservatives that Middle Eastern instability posed a direct threat to US security, and that decisive military action was needed to exert control over potentially dangerous nations in the region. While Afghanistan, as the designated origin of the attacks, became the immediate target, the desire to link the Iraq question to the terrorism problem began to intensify within policy circles. In February 2002, for instance, US Secretary of Defence Donald Rumsfeld claimed that the US knew that Iraq possessed chemical and biological WMDs, and insinuated that terrorist groups may seek to obtain such weapons from Iraq (US Department of Defense, 2002). The terrorist threat, in this respect, was used as a platform for advocates of intervention to sell a war against Iraq to the US population.

The primary legal justifications for invading Iraq were related to US contentions that the Ba'ath regime possessed and would use WMDs, and the espoused links between Iraq and al-Qaeda.[4] The moral justification for the war, however, was couched in the language of rights and the rights violations inflicted on ordinary Iraqis by the Ba'ath regime. For instance, in an address to the UN General Assembly in September 2002, President George W. Bush presented a typically emotive account of Ba'athist violations:

> Last year, the UN Commission on Human Rights found that Iraq continues to commit extremely grave violations of human rights, and that the regime's repression is all pervasive. Tens of thousands of political opponents and ordinary citizens have been subjected to arbitrary arrest and imprisonment, summary execution, and torture by beating and burning, electric shock, starvation, mutilation, and rape. Wives are tortured in front of their husbands, children in the presence of their parents – and all of these horrors concealed from the world by the apparatus of a totalitarian state.[5]

As you can see, Bush's articulation of Ba'athist oppression strikes a similar tone to Walzer's depiction of acts the shock the moral conscience of humankind: violence is presented as a morally necessary response to excessive rights violations. Despite similarities in the language employed by the Bush Administration, it must be acknowledged that Walzer consistently rejected the idea that the US invasion constituted a just war (2005, 2006b, 2008, 2012). For Walzer (2005), a continuation of the sanctions regime and limited strikes against military targets would be more useful in helping Iraqis to wrestle self-determination back from their Ba'athist oppressors. In addition, Walzer definitively asserts that regime change, in itself, is never an acceptable justification for war (2006b: 105) – thus he considers the US invasion of Iraq to be unjust and 'potentially' an act of aggression. Nevertheless, Walzer is hesitant to condemn US actions and is keen to emphasise the ability of unjust wars to achieve just outcomes (2012: 44). Walzer, in this respect, wants to convey an understanding of ethical responsibility in which doing the morally wrong thing in the first instance does not preclude the achievement of a morally right outcome: even though the US war was unjust, it could still produce a just Iraqi society. Hence Walzer's understanding of *jus post bellum* (just resolution to war) in this context can tell us a lot about his understanding of the US invasion.

Just and Unjust Wars and Walzer's broader discussions of war are particularly inattentive to the question of post-war justice. Indeed, Walzer acknowledges that his conception of *jus post bellum* has proven ineffective in explaining the problems that arose during contemporary humanitarian interventions after the end of the Cold War (2005: xiii). The main reason for this, Walzer (2012) explains, is that his understanding of just resolution is intimately tied to *jus ad bellum* (just cause). Walzer's theory of *jus ad bellum* implies that war can be justified only when a community is faced with an aggressor who threatens their internal peace and stability. As we have discussed in the previous chapter, this understanding of justified violence presupposes that a community of members separated from the outside defines just and peaceful existence. In Walzer's terms, the world is at peace when it is divided into established and widely accepted borders, and justice is challenged if an aggressor disrupts the arrangement (2012: 35). Walzer's central argument is that war is a crime because the aggressor violently disrupts the established conditions of peace within a bounded community. His image of just resolution logically follows from this ideal of internal peace; principally, the idea that the primary aim of just resolution is to restore the conditions of internal peace: to restore self-determination (2006a: 121–122). In this sense, a just resolution is a resolution that protects the common life and shared meanings that members of a community have built together. Further to this, Walzer stresses that those intervening against aggressors should not aim to change social meanings or even redress unjust distributions. In fact, the sole aim of just resolution is to restore the political foundations through which members of a community can continue to negotiate their shared resistance (2012: 36). In short, a just resolution is one that preserves the conditions for communal self-determination.

While conventional resolutions are relatively straightforward, Walzer (2006b, 2008, 2012) argues that humanitarian intervention demands a more nuanced understanding of just resolution. Most immediately, the disruption of self-determination comes from inside the community and not from an external aggressor. In cases of intervention, Walzer (2006a) reminds us that the state has turned so savagely on its own people that it is, in effect, suppressing the possibility of self-determination. Following this, he suggests that the secondary goal of humanitarian intervention demands the creation of the minimal structures necessary for self-determination to flourish:

> In the case of humanitarian intervention, *jus post bellum* involves the creation of a new regime, which is, minimally, nonmurderous. And it is more than likely that the creation of a new regime will require some period, perhaps an extended period, of military occupation.
>
> (Walzer 2012: 38–39, original italics)

Walzer reconceptualises intervention resolutions as the external cultivation of the minimal requirements of self-determination. While this in certain respects re-articulates Walzer's image of community – the ideal of protecting the ability of a community to create and sustain a common life – it also suggests that creating the conditions for self-determination is sometimes the responsibility of strangers. In contrast to ordinary political communities that foster self-determination auto-affectively, Walzer argues that post-intervention societies require external assistance in creating their maximal world. Walzer explains that external help is required because internal animosities often mitigate against nonviolent politics: 'A devastated country in which the killers and the people they tried to kill (and whose relatives they did kill) live side by side is not a likely setting for democratic deliberation, popular engagement, and nonviolent opposition' (2008: 351). As such, the just resolution to an intervention is not conceived in terms of a return to the former conditions of peace. Instead, the intervening force(s) is responsible for ensuring that it leaves behind a system of governance that is capable of promoting self-determination and protecting ordinary people from violence.

Walzer's acknowledgment that external actions are sometimes necessary to help a community cultivate self-determination marks an important departure in his understanding of morality. Throughout Walzer's entire *oeuvre* he persistently asserts that border crossings that attempt to alter internal meanings and values exemplify aggression, and that the external transformation of the inside by the outside is morally wrong. When it comes to intervention, however, Walzer suggests that external transformation of internal politics is the morally right thing to do. He attempts to reconcile the tension between the necessity for intervention and the necessity to preserve self-determination via an appeal to his minimal/maximal (universal/particular) dichotomy. Walzer contends that *post bellum* responsibilities do not amount to the right to impose external rule, or political ideology, or a specific political system on liberated people; intervention should not aim to construct a full-blooded maximal life. Instead, the intervening force(s) must endeavour to found a model of self-determination that adheres to minimal morality while remaining attentive to local desires:

> The intervening state can't then impose its version of a just politics without regard to their version . . . local understanding of political legitimacy is a critical constraint on what just warriors can attempt. *But it isn't an absolute constraint.*
> (Walzer 2012: 43, italics mine)

Walzer signals an ideal of post-war justice in which the intervening strangers must take account of local understandings (maximalism), but ultimately have some scope to overturn particular local values in the name of universal norms (minimalism). For example, Walzer defends the US imposition of a democratic constitution in Japan after the Second World War, even though it directly challenged existing Japanese customs (2012: 43). Walzer describes this ideal of intervention justice in terms of the interplay between local norms and minimal rights. In this sense, Walzer is pointing toward an understanding of post-intervention justice in which self-determination is a negotiation between existing maximal values and minimalism. This conception of justice, nevertheless, marks an important shift in Walzer's theory of morality. In contrast to the image of minimalism founded within the intersection of maximal moralities as outlined in his communitarian writings, Walzer is now implying that the creation of minimal foundations is sometimes necessary before a genuine political community can emerge. The outside intervenes in order to establish the conditions through which a shared life can be fostered and hence, minimalism founds the possibility of authentic maximalism.

Nonetheless, we must acknowledge that Walzer is only advocating this form of communal construction in cases where, he claims, existing governments have violently suppressed maximal life. What is more important, in understanding Walzer's reversal, is the distinction between minimal structures and their maximal articulations. Although Walzer (2012) suggests that strangers can create political structures that defend minimal values, he does not necessarily believe that the imposition of minimalism undermines or alters the ability of a community to construct their own maximal life. In this sense, the imposition of minimalism constitutes the imposition of universal structures that are conducive to the development of genuine maximal life. In other words, the cultivation of minimal structures is equivalent to building the universal container that self-determination needs to grow within. For example, Walzer suggests that interventions should aim toward democracy because it is inclusive of all members of the community and is the form of political regime least likely to turn on its own people (2012: 44). Yet Walzer stresses that the argument he is making is distinct from cosmopolitanism, because minimal democratic structures still allow local populations to substantiate their particular interpretation of what democracy means to them:

> Struggles for democratization, whatever help they receive from outsiders, are always local struggles. Their protagonists do not aim at the triumph of cosmopolitan principles around the world. They want a state of their own, in the literal sense of that term – a state governed by the people who live in it, devoted to their welfare.
>
> (Walzer 2008: 355)

What is important to note is Walzer's belief that the minimal structures necessary for a just resolution to an intervention do not suppress the ability of locals to freely create a shared life. For Walzer, a just intervention creates the minimal structures through which authentic maximal life is produced and protected without altering its substantive content. Just resolution, therefore, remains rooted in the ideal that war can leave behind a community whose shared values, common life and capacity for authentic community building have not been distorted by the intervention.

In many respects the Bush Administration's overarching conception of just resolution in Iraq is aligned with Walzer's, at least in terms of rhetoric. Specifically, the primary pillars of freedom and democracy evident within the Administration's rhetoric closely correlate to Walzer's conception of minimal self-determination. For example, in a speech to the American Enterprise Institute on 26 February 2003, US President George W. Bush stated:

> There was a time when many said that the cultures of Japan and Germany were incapable of sustaining democratic values. Well, they were wrong. Some say the same of Iraq today. They are mistaken. The nation of Iraq, with

its proud heritage, abundant resources and skilled and educated people, is fully capable of moving toward democracy and living in freedom.

(Bush 2003)[6]

In Bush's speech we see a clear articulation of Walzer's conception of post-war justice, an intervention aimed toward sustaining democracy and legitimate government. In fact, Walzer tentatively endorsed the image of post-war justice that the Bush Administration envisioned for Iraq:

> They [the things America is required to provide for the Iraqi people] include self-determination, popular legitimacy, civil rights and the idea of a common good. We want wars to end with governments in power in the defeated states that are chosen by the people they rule – or at least recognised by them as legitimate – and that are visibly committed to the welfare of those same people (all of them). We want minorities protected against persecution, neighbouring states protected against aggression, the poorest of the people protected against destitution and starvation. In Iraq, we have (officially) set our sights even higher than this, on a fully democratic and federalist Iraq, but postwar justice is probably best understood in a minimalist way.
>
> (Walzer 2005: 164)

Walzer maintains that the Bush Administration perhaps aimed too high; but nevertheless, he supports the ideal of helping the Iraqi community move toward some form of inclusive socially attentive self-determination.[7] In this respect, the professed goals of the Bush Administration's invasion of Iraq echo Walzer's overarching conception of post-war justice. Hence, the discourse of rights and universalism that Walzer believes to underpin minimal morality was implicitly tied to the ways in which the Bush Administration sold the war to various publics. In turn, Walzer argued that although the war was unjust, it could be potentially justified by creating the minimal foundations necessary for a just Iraqi society.

Despite the emphasis placed upon WMDs and alleged links to terrorism, the proposed resolution for the Iraq war was presented in explicitly humanitarian terms: US military force would secure democracy and freedom for the Iraqi people. Iraq, then, provides an important fusion between the neoconservative desire to exert regional influence over the Middle East and the humanitarian goal of saving ordinary Iraqis from abuse. In this way, the humanitarian desire to protect rights, in particular the right to freedom, became mobilised within US justifications for the proposed invasion. Elshtain (2002), for example, argues that the US had a moral responsibility to defend the Iraqi victims of Ba'athist oppression, and this could only be achieved through military force.[8] Rights discourse, in this way, represented a key dimension of how the war was presented: intervention would protect US citizens from the threat of terrorism and WMDs while simultaneously liberating Iraqis from Saddam Hussein's tyranny. US justifications for the invasion hinged upon defending rights and pre-emptively quashing threats. The political plan for Iraq, however, hinged upon strategically placing Iraqi exiles in key positions within the new political infrastructure. In the years leading up to the invasion, the Bush Administration had courted a pool of Iraqi exiles opposed to the Ba'ath regime.[9] The exiles can be broadly categorised as liberal secularists who endorsed Western-style democratic government. Many had been educated in the West and most had lived in Western nations for years, even decades, prior to the invasion. The US believed that Iraqi democracy could be unified around a distinctively liberal and secular ethos that countered any potential sectarian or religious divisions. In this respect, the Bush Administration was sensitive to the fact that democratic governance would favour Iraq's Shi'a majority that could potentially align itself with Iran[10] and its anti-US stance (Napoleoni 2005: 160). As such, the US ensured that exiles from religiously inclined Shi'a political parties were largely excluded from the process (Allawi 2007: 69). With the invasion geared toward the instillation of a secular democracy, a liberal regime of rights and free-market economics, the exiles provided the perfect foil to US force: a vetted selection of Iraqis predisposed to the type of society that the US hoped to create in post-Ba'ath Iraq.

The Bush Administration largely viewed Iraq as a blank slate ripe for liberal democratic values, and there was never any question that Iraqis might resent their 'liberation' or oppose these 'universal' values. Shadid sums up the mythical image of post-war Iraq that permeated US decision-making prior to the invasion: 'With Saddam gone, jubilant Iraqis would embrace those who had freed them. Together with the exiles, the Americans would create an outpost of democracy and prosperity in a region with little of either' (2006: 157). The mythos of pre-invasion Iraq was, in part, generated through an unwavering belief in the universal appeal of the values underpinning liberal democracy. The Bush Administration simply could not conceive a situation in which any people would reject the form of society it hoped to create. Importantly, the US firmly believed that the various ethno-sectarian facets of Iraqi society could be unified around the universal values embedded in liberal democracy. In President Bush's words, 'We're committed to the goal of a unified Iraq, with democratic institutions of which members of all ethnic and religious groups are treated with dignity and respect'.[11] The Bush Administration's political ideals are closely related to Walzer's understanding of the role that minimal morality plays in post-intervention societies. While Walzer would undoubtedly challenge the belief that liberal secularism represents a minimal code, the structural ideal remains identical: if you want to create a unified political community in a divided nation, you must build the minimal/universal foundations necessary to facilitate the emergence of a genuinely shared maximal world. US post-invasion policy was guided by this belief, and the type of Iraq that the US sought to build started from the assumption that liberal democracy was a universal good embraced by all rational people.

Occupation and legal authority

The US invasion of Iraq commenced on 19 March 2003 and major combat operations were declared to have ceased by 1 May 2003. Iraq, in this respect, appeared to represent a relatively minor war: the US entered Iraq with a large ground force and rapidly moved through the country, capturing major cities, routing Iraqi troops and removing Ba'athist leaders.[12] Iraq provided an easy victory for the US in terms of conventional warfare. The difficult task, however, was always going to be rebuilding Iraqi society and infrastructure and restoring democracy. The US plan for Iraqi society was broadly reflected in two policy initiatives. First, US political administration of Iraq was proposed as a means to cultivate institutions and legal infrastructures compatible with democratic politics; and second, US troops were committed to Iraq in order to foster long-term peace. These initiatives were enacted through the formation of the Coalition Provisional Authority (CPA) on 21 April 2003 and the UN ratification of US Occupier Status in Iraq on 22 May 2003. The CPA aimed to develop Iraqi legislation that could pave the way for the emergence of a liberal democratic society – the primary goal being the creation of a new constitution that would legally enshrine Iraqi democracy (Allawi 2007: 191–193). In turn, the UN resolution legalising US Occupier Status clarified the rights and duties of US administration and troops in Iraq, and gave the CPA access to Iraqi funds deemed necessary to finance the transition to democracy (Karon 2003). These dual initiatives comprised the main legal infrastructures through which the US attempted to cultivate Iraqi democracy and protect ordinary Iraqis from rights violations.

The CPA represented a clear attempt by the US to formally institute their legislative authority over the governance of Iraq. Paul Bremer (head of the CPA) declared his intention to shape Iraq's future: 'We will dominate the scene . . . and we will impose our will on this country' (Gregory 2004: 217). The CPA quickly asserted itself as the primary source of Iraq legislation (Chandrasekaran 2008: 69) and formally stated that elections would not take place until the CPA had drafted a constitution (Allawi 2007: 190–193).[13] In other words, the CPA was designed to lay the foundations for a new Iraqi democracy. The CPA aimed to foster mass cultural, political and economic change in Iraq in a number of ways (Cockburn 2007: 71). CPA changes ranged from attempting to promote secularism by symbolically cutting off Islamic prayer from national television (Chandrasekaran 2008: 148), to fostering a neoliberal economic system by restructuring and privatising Iraq's economy (Gregory 2004: 244–245), to the cultural overhaul of Iraq's education system via the implementation of a Westernised curriculum (Allawi 2007: 383). The CPA's early actions signified the clearest articulation of the US conception of Iraq's future: democracy and freedom would be bestowed upon Iraqi people. Yet these goods would be withheld until such time as the US could ensure that Iraq became the 'right' kind of democracy. In Walzer's terms, US actions in Iraq began to conflate the minimal and maximal worlds by treating a particular understanding of democracy as if it were universal. This was not, however, an accident, or necessarily the result of misguided US policy and beliefs. Rather, as will be explained, the US attempt to rebuild Iraqi democracy illustrates the tension between the desire to cultivate minimal values and the fact that minimalism is itself unrepresentable in a minimal way.

Cultivating a political system that reflected liberal democratic values within a largely unknown cultural entity presented the US with a difficult task. Even the exiles that the US had relied upon for cultural insight prior to the invasion were surprised by the extent to which Iraqi society had changed under the post-Gulf War sanctions regime (Cockburn 2007: 93). As such, the US began to embrace the colonial strategy of structuring Iraqi politics along sectarian lines (Gregory 2004: 230). The clearest example of this strategy was evident in the formation of the Iraqi Governing Council (IGC), a CPA-appointed body, which was instituted to allow Iraqis to have an input with regard to CPA policy. The CPA appointed a total of twenty-five Iraqis to the IGC: thirteen Shi'a, five Kurds, five Sunni, one Turkoman and one Christian (Jamail 2008: 182). The US strategy followed from the logic that each ethno-sectarian faction in Iraq had a distinct set of cultural understandings and values. Following this assumption, the US decided that the fairest form of political representation was to assign positions based on ethno-sectarian identity in accordance to population demographics. The Shi'a population, for instance, was the largest demographic and, therefore, should have the highest representation. Although the vast majority of appointees were selected from the collective of exiles, the strategy nonetheless began to magnify the importance of ethnicity and sect as an identification marker. The political weight that the US placed on ethno-sectarian lineage represented a marked difference in the role that identity played in Iraqi society. In Baghdad, for example, it was often considered ill-mannered to even inquire about someone's ethnic identity (Shadid 2006: 104). Shadid explains the dynamics of Iraqi identity in terms of overlapping and intertwined ethno-sectarian and tribal bonds:

> The distinctions among Sunni, Shiite, and Kurd often mean less in Iraq than they do in the West . . . Tribes, sometimes included both Sunnis and Shiites still play a powerful, even resurgent role in Iraq. Most Kurds are Sunni, but some count themselves Shiites. Among the Shiites themselves, there are gradations in identity: between the secular and the religious; across a loose caste system built around descendants of the prophet Mohammed's family; between the modern educated and those still moulded by the durable but traditional scholastic institutions . . . There is no less diversity among Sunnis.
>
> (Shadid 2006: 103)

Shadid outlines the more complex relationship between familial, tribal, regional, ethnic and national loyalties that underscored pre-occupation Iraqi identity. In fact, tribal bonds began to grow in importance during the final years of the Ba'ath regime (Allawi 2007: 14). For example, Hussein became increasingly reliant upon support from his own tribal community in Tikrit as the 1990s progressed and bestowed increased powers upon 'his' people (Dawisha 2009: 236). In other words, the ethno-sectarian divide was not always conceived by Iraqis as their main identity marker. In cases where ethno-sectarian ties became central to identity, this was usually in response to violence, repression and marginalisation. Kurdish identities in northern Iraq and Shi'a identities in the Mesopotamian Marshes, for instance, were solidified in opposition to Ba'athist brutality during the 1980s and 1990s (Cockburn 2007). This is not to imply that ethno-sectarian markers were not important in regard to pre-occupation Iraqi identity. Instead, I want to emphasise that the formal categorisation of Iraqis in terms of their ethno-sectarian heritage, and explicitly linking these markers to political representation, was a construct devised under US administration. In turn, this formal construct helped cement a particular understanding of Iraqi identity. According to Saad Jawad, a professor of political science at Baghdad University, Iraqi perceptions of identity were transformed because the US 'made a point of categorising people as Sunni or Shiite or Kurd' (cited in Chandrasekaran 2008: 218). That is, by restructuring politics explicitly along the lines of ethno-sectarian identity, the US began to recontextualise Iraqi conceptions of politics in terms of ethno-sectarian competition.

The UN resolution granting the coalition forces occupier status was passed on 22 May 2003, and this legal declaration of occupier status proved to be another major event in US relations with Iraqi society. The logic behind this move was to reintegrate the UN in Iraq's future, and to legally clarify the rights and responsibilities of US authority in Iraq. In this sense, the move toward occupier status sought to draw a line under the controversy surrounding the legality of the war, and bolster the chances of a broader international reconstruction effort. However, the term 'occupation', in an Iraqi context, is itself loaded with political, historical and colonial connotations that began to redefine public perceptions of what US intervention meant. Shadid explains that the term *ihtilal* [occupation] denotes an unequal relationship, a weaker power submitting to the will of a stronger one; and this changed Iraqi perceptions of what the fall of Ba'athism symbolised (2006: 237). Instead of seeing the intervention as a chance for Iraqis to take control of their nation, occupation suggested that the US were now in charge of Iraq's destiny: one autocratic regime had been replaced by another, foreign, one. The declaration of occupier status points toward another disconnect between US imaginings of Iraq and the society they faced. Pre-invasion, the US viewed the Iraqi state as an extension of Saddam Hussein and the Ba'athist regime: Iraq was Hussein and defeating Hussein would be welcomed by ordinary Iraqis. Yet under Ba'athism, Iraq retained semblances of more multi-faceted identities and conceptions of nationalism in which the regime was one aspect of a vast cultural and historical assemblage. This fractured and uneasy nationalism is illustrated in the US's failed attempt to replace the Iraqi flag. The US believed that the flag represented Hussein and that Iraqis would relish the chance to get rid of it (in fact, Kurdish communities did). However, the flag pre-existed the Ba'ath regime and many Iraqis felt that it embodied an aspect of their shared identity that Ba'athism had not destroyed. In the words of a Baghdad shopkeeper, '[t]he flag is not Saddam's flag. It was there before Saddam and it represents Iraq as a country. The whole world knows Iraq by its flag' (Cockburn 2007: 146). The contestation of the Iraqi flag underlines the tension between the US's image of Iraq as a blank slate upon which liberal democracy could be inscribed, and the understanding of Iraq as a historical entity. Ba'athism and its policies may have dominated Iraqi society for a number of decades, but this did not necessarily mean that Iraqis felt that the defeat of the regime erased other elements of Iraq's history and culture. By legally defining Iraq as a conquered nation and withholding elections until the US had instituted a new legal framework for Iraqi society, the US alienated many factions of Iraqi society from the political process, and encouraged political groupings to seek influence through alternative channels.

Post-war security

One of the most immediately visible effects of the overthrow of the Ba'ath regime was large-scale looting and civil unrest. Banks, businesses, museums and, perhaps most importantly, government ministries were the primary targets for looters.[14] Shadid, living in Baghdad during the invasion, described the looting in terms of the sacking of a medieval city (2006: 140). Indeed, many Iraqis began to understand the looting through the historical frame of the sacking of Mesopotamia by the Mongols in the thirteenth century (Napoleoni 2005: 148). This image intensified as looters turned their attention to Iraq's national museums and their cultural treasures, many dating from the Mesopotamian era. While US Secretary of State Colin Powell sombrely reflected that the US Military had not anticipated such a widespread collapse of Iraqi civil society (Dawisha 2009: 243), Rumsfeld, rather audaciously, dismissed concerns by stating that 'freedom's [*sic*] untidy' (cited in Shadid 2006: 157). Crucially, the US military leaders gave their troops no orders to stop the looting (Filkins 2009: 98). In fact, Iraqis reported seeing US soldiers taking photographs of looters like some form of perverse tourism and, in other cases, actively cheering looters on (Cockburn 2007: 75). Again, the US response to looting indicated a misunderstanding of the relationship between Ba'athist rule and the Iraqi state. The Bush Administration viewed looting as a popular reactionary outpouring of anger and opportunism from an impoverished people who had been brutalised and oppressed for the latter half of the last century. In other words, looting represented ordinary Iraqis tearing down the cultural and economic vestiges of the old repressive order. However, looting also contained traces of more organised attempts to destabilise the occupation. For example, looting was used as a cover for the sabotage of oil pipelines, and government offices were targeted as a means to destroy Ba'ath files, records and databases. Perhaps most importantly, the looting of numerous Iraqi Army arms caches provided a source of military firepower that would become incremental to resistance fighters[15] and various militias in the post-war period (Allawi 2007: 140). The US misunderstanding of the political dimensions of looting, therefore, helped to provide the material and ideational kindling necessary to ignite a sustained resistance to the occupation.

At the level of ordinary Iraqi people, the prevailing violence and criminality that accompanied looting began to have profound effects on daily living. Shamselin Mansour's account of what daily life had descended to in post-war Baghdad helps to illustrate the fear and anger felt by Iraqis coming to terms with the new social order:

> We live as many as forty-two people in a house and do not have the money to buy even a small generator. Without light at night it is easy with gangs of

thieves with guns to take over our streets, and shooting keeps us awake. If we try to protect ourselves with guns, the Americans arrest us.

(cited in Cockburn 2007: 104)

Shamselin's account draws together a number of key problems faced by Iraqis during the early days of the occupation. The consequences of infrastructural decay and social collapse became increasingly visible during the first years of the occupation; for example, unemployment stood at 70 per cent after the invasion (Cockburn 2007: 16), only 37 per cent of water was being treated (Gregory 2004: 222) and in some areas of Baghdad, raw sewage spilled onto the street, spreading a myriad of diseases including hepatitis E (Jamail 2008: 197). In addition to the economic and health risks, a clear vacuum of direct authority was becoming increasingly apparent. While the US was focusing on organising Iraqi democracy at a national level, apart from the visible presence of US troops, there were limited efforts made to provide day-to-day security, or to engage with local political grievances. This vacuum was quickly filled by various armed groups who saw an opportunity to garner power and shape the post-war political environment at a sub-national level.

The Sadrist movement provides an illustrative example of how this form of localised authority began to take hold. The movement was led by the young Shi'a cleric Muqtada al-Sadr, the son of a Grand Ayatollah murdered by Ba'athists in 1999.[16] The Sadrists were based in Saddam City, a sprawling Baghdad slum with a mass populace suffering from appalling social conditions. In some respects the Sadrist military wing, the Mehdi militia, represented the first boots on the ground in post-war Baghdad. In the midst of the post-war chaos, Sadrists seized the initiative by symbolically renaming their slum 'Sadr City' in honour of their martyred Ayatollah; Mehdi troops took direct responsibility for public security and service provision: they organised armed Mehdi brigades to protect local mosques, businesses and hospitals from looters (Gregory 2004: 226). The Mehdi instituted itself as a local police force (Jamail 2008: 119–120) and the movement began to cultivate independent infrastructures for local service provision through Shi'a mosques (Napoleoni 2005: 137). Sadrists even presented themselves to US troops as mediators through which aid could be distributed (Fick 2007: 341–443) and appealed for military resources to root out former Ba'athists (Wright 2005: 414). The hands-on response of the Sadrists stood in contrast to that of US troops who watched the post-war carnage unfold as they awaited superior orders. Soon, locals began to trust the Sadrists as the primary security force in Sadr City.

While direct militia activism announced Sadrist presence, their sustained appeal was underpinned by religious authority. In the immediate aftermath of the invasion, Sadrist clerics issued a *fatwa* – a binding religious decree in the Shi'a faith – ordering all Shi'a to ignore US authority and to fight against Western cultural corruption of Islamic values (Cole 2003: 554). Sadrists represented a direct challenge to the US's proposed plan for Iraq by explicitly rejecting US authority on the basis of the perceived antagonism between Western norms and religious precepts. The Sadrists, in other words, openly rejected the presupposed universal values espoused by the Bush Administration because they viewed them as the forced imposition of explicitly Western values. In Walzer's terms, the Sadrists argued that US minimalism was, in fact, a robust form of threatening maximalism. Sadr's legacy of personal loss became an integral component of the movement's political legitimacy and his stature grew as a symbol of those who had suffered under Saddam Hussein.[17] Importantly, the Sadrist movement encapsulated an increased willingness on the part of religious authority to step into the political arena. In the words of Sadrist cleric Ali Shawki, '[t]he religious man is not confined to the pulpit . . . He can act as a military, political, social, and spiritual leader' (cited in Shadid 2006: 189). Shadid stresses that the Sadrist conception of political ideology did not expand beyond a somewhat hazy notion of the centrality of Islamic values (2006: 208). Nevertheless, the movement's overt political ambitions marked a departure from traditional Shi'a conceptions of religious authority (Cole 2003), and the movement differentiated itself from other Shi'a voices through its direct and vocal opposition to the occupation (Goldwin 2012: 450). Interestingly, during the early stages of the occupation the movement attempted to foster a broad Iraqi nationalist opposition to the US by allying with Sunni resistance fighters (Allawi 2007: 168). In this respect, the Sadrists were interested in creating an Islamic state in opposition to the US occupation, rather than an explicitly Shi'a political authority. In total, the Sadrists embodied a grassroots localised Shi'a political response that opposed the occupation through an explicitly Islamic and nationalistic framework.

The Sadrists, however, were not unique in assuming responsibility for security and policing in post-war Iraq. Kurdish militias in the north, Shi'a rebels in the south and various Sunni groups soon began to stamp their authority in various localities. Ultimately, Shi'a and Kurdish militias were incorporated within the state policing apparatus, thereby legitimising their status as security providers. While US troops attempted to assume more direct control over policing duties once the looting had subsided, the damage to Iraqi perceptions had been done: the US had stood by and watched as criminals tore their country apart, and only local groups had stood up and tried to defend the ordinary people. This signified a new chapter in Iraqi politics in which religiously and ethnically defined security forces began to emerge as the primary source of local authority. However, the dismantling of Iraq's military and governmental infrastructure under the US policy of de-Ba'athification ultimately solidified this perception, and stimulated the emergence of sustained resistance to the occupation.

De-Ba'athification and Sunni resistance

During the early stages of the occupation, the US characterised Iraqi resistance under two distinct headings: Ba'athists seeking to return Saddam Hussein to power; and foreign jihadists fighting a broader war against the West. In 2003, for example, Rumsfeld described the resistance as Ba'ath party 'dead-enders' (cited in Kelly 2003) and Bremer claimed that al-Qaeda was leading the resistance (Allam 2003). The central motif in the US depiction was that resistance was coming from the relics of Iraq's despotic past and from foreign religious fanatics. In other words, the resistance did not represent the views of ordinary Iraqis. In some respects, the US picture was not entirely inaccurate: there is some evidence that suggests former Ba'ath security forces were engaging in resistance activities and sabotage (Ricks 2007), and foreign fighters were crossing Iraq's borders in relatively large numbers (Napoleoni 2005). The US depiction, however, misrepresented resistance and detracted from Sunni political discontent directly related to US actions. Carter Malkasian argues that Sunni resistance did not represent a coherent or unified movement, nor did it adhere to a singular ideology or political goal (2006: 271). In this sense, Sunni resistance comprised of a multitude of local, often loosely connected, groups who, for various reasons, opposed the occupation: '. . . initially the insurgents emerged . . . from personal, tribal, sectarian, Islamic, Arab and nationalist resentment at the myriad of humiliations of the fact and conduct of the occupation' (Herring and Rangwala 2006: 167). To understand why Sunnis resisted the occupation, it is therefore important to look at a few examples of how US policies impacted on Sunni communities.

Ahmed Hashim explains that while Sunnis enjoyed a privileged position under Ba'ath rule, many Sunnis resented the regime and some Sunni communities (for example, Sunnis in Samarra) actively opposed it (2003: 3–4). In fact, Dawisha (2009) argues that during the 1990s the Ba'ath began to cement privilege more directly within Saddam Hussein's tribal community in Tikrit, thereby alienating the majority of Iraq's Sunni population from Ba'athism's inner circles. However, we cannot ignore the fact that the invasion marked the end of Sunni dominance over Iraqi politics which had lasted since the British occupation in the early twentieth century. The Sunni transition to a minority role was, perhaps, always going to be a thorny issue in the post-war era. Nonetheless, the invasion precipitated events that exacerbated Sunni perceptions that they were being unfairly targeted because of their ethnic association with Ba'athism. Post-war looting provides a good example of how this perception was cultivated. Wealthy Iraqis were by and large Sunni and therefore were more likely to own businesses and the most expensive homes. As such, when looting broke out, Sunni homes and businesses were the primary targets – not necessarily because the owners were Sunni, but because they were the most economically attractive to looters (Napoleoni 2005: 141). While the Bush Administration was happy to brush off looting as liberated Iraqis 'blowing off steam' (cited in Hoyt and Palatella 2007: 31), the refusal of US troops to stop the looting was interpreted as a direct attack on Sunni communities. Sunnis felt that US troops stood aside because Sunnis had supported Hussein and deserved to be punished. Despite the perceived injustices, looting by itself did not cement Sunni marginalisation. Instead, US plans to radically restructure Iraq's public sector and military became the focal point of discontent.

US justifications for the war were predicated upon a definite rejection of Ba'athism, and the Bush Administration promised to purge Ba'ath party members from Iraq's public institutions. This ideal is related to Walzer's conception of post-intervention justice. Walzer argues that the aim of intervention is to facilitate the emergence of minimally 'nonmurderous' regimes, i.e. regimes that will not turn violently on ordinary people. Therefore, we can assume that replacing the previous 'murderous' regime is integral to Walzer's conception of justice. In the context of Iraq, Walzer (2005) is clear that the deposal of the Ba'ath regime would help bring justice to Iraqi society. The proposed link between justice and de-Ba'athification meant that overhauling the existing civil service infrastructures soon became one of the CPA's most pressing concerns. Yet de-Ba'athification, in practical terms, was tantamount to the radical transformation of Iraq's economy and society. Under the 1990s sanctions regime and its restrictions on international trade, Iraq's economy had become increasingly centralised. By 2003, Iraq's public sector was one of the few remaining sources of reliable legal income (Cockburn 2007: 16). The problem with de-Ba'athification was that public sector employees were often required to join the Ba'ath party or face dismissal. Ba'ath party members, in this respect, were often Iraqis who held no political allegiance to Ba'athist ideology but who wanted to secure a stable income in a country with scant employment opportunities. The US acknowledged this and attempted to avoid unfairly punishing public sector workers by restricting de-Ba'athification to the four upper ranks of party membership, claiming that this would limit dismissals to 20,000 staunch Ba'athists (Ricks 2007: 160). The US rolled out de-Ba'athification through CPA Order Number One on 16 May 2003. However, they misjudged the implications of the order and a total of 85,000 to 100,000 people, including 40,000 ordinary teachers, were ultimately dismissed (Pfiffner 2010: 79). The order affected Sunni communities to a far greater extent than Shi'a and Kurdish ones, because Sunnis were more likely to be employed by the state and to be members of the Ba'ath party. In some Sunni areas the order crippled public infrastructures; for example, local schools had to close in isolated north-western Iraqi villages and towns because the majority of teachers were dismissed by the order (Chandrasekaran 2008: 81).

De-Ba'athification was followed on 23 May 2003 by CPA Order Number Two, which disbanded the Iraqi security forces. The order was, once again, underpinned by the ideal of purging Iraq of Ba'athism. In the case of security forces, the US was steadfast in its belief that Iraq's military institutions needed to be completely dismantled. In the eyes of the Bush Administration, the intervention was necessary because of the brutal nature of Hussein's regime and the security forces were the primary instruments of this brutality. As such, the dissolution of security forces was seen as a necessary and symbolic move away from Iraq's former identity. In the words of a senior CPA official, the US wanted to 'show the Iraqi people that the Saddam regime is gone and will never return' (cited in Arraf 2003). Nevertheless, the second order had dramatic implications for Iraqi society: it dissolved the Army, the Ministry of the Interior, which included police and domestic security forces, and presidential security troops. In total, 700,000 Iraqis were dismissed from military jobs (Ricks 2007: 162). Again, Sunni communities were most affected; for example, 100,000 ordinary soldiers and 1,100 officers lived in the largest Sunni city, Mosul (Hashim 2003: 7). In Iraq's devastated economy, the effect of disbanding the security forces was crushing, and close to a million security personnel were left without any income or any means to support their families. Shadid argues that targeting Iraq's security forces created a reservoir of angry and humiliated Sunni men who possessed some degree of military training (2006: 181). De-Ba'athification, in this way, set the stage for the sustained resistance movement that emerged in Sunni regions during the following years. Soon the Sunni Triangle, an area in north-western Iraqi dominated by Sunni populations, became the centre of resistance operations (Foot *et al.* 2004: 57). In more general terms, the dissolution of the Army, one of Iraq's oldest institutions, convinced Iraqis that they were indeed living in a fallen state under foreign control.

The opening months of the occupation resulted in a dramatic shift in the status and societal position of Sunni populations. The consequences of looting and, more importantly, de-Ba'athification, transformed some of the most economically stable communities in Iraq into collectives united through poverty and exclusion. This, in turn, began to cement Sunni perceptions that they were being written out of Iraq's future. As Hashim contends, Sunnis felt that the reversal of their socio-political fortunes, and their inadequate political representation in the IGC, was an attempt to punish Sunni populations for Hussein's crimes on the sole basis of their ethno-sectarian identity (2003: 4). The post-invasion plight of Sunni populations highlights the inadequacy of Walzer's conception of regime change. Walzer argues that regime change means getting rid of those responsible for state brutality, which implies a rather straightforward deposal of state leaders, as mirrored by the Bush Administration's pack of playing cards representing the main Ba'ath offenders. The political reconstruction of Iraq, however, illustrates that the more problematic decisions revolve around ordinary citizens who were employed in wider state apparatuses, both administrative and military. Walzer maintains that people should not be punished on the basis of who they are, but on the basis of their individual actions (2006a: 135). Iraq illustrates why it is difficult to ascertain individual culpability in regard to state crimes: in a regime shrouded in secrecy it was impossible to know who had committed what acts, and why. To punish state employees, in any responsible way, would entail a prolonged and expensive judicial process – and the US did not even entertain this possibility. On the other hand, de-Ba'athification highlights how generalised approaches to post-war justice sow the seeds of long-term opposition to any new regime. What is important is that regime change never simply entails the deposal of a 'murderous' state. It is simultaneously bound up in the larger question of how we justly determine who is responsible for state brutalities and the ethical implications of determinations of guilt.

Religious authority and the Iraqi constitution

Religious identity in Iraq was carefully managed under the Ba'ath regime, ensuring that religious authority did not stray into mainstream political life. The management of religion was primarily motivated by Ba'athist fears that Shi'a religious leaders could pose a direct threat to its rule, a fear that was exacerbated by the 1979

Iranian revolution. The US, however, emphasised their aim to free Iraqis from tyranny and, despite similar fears about Shi'a religious leaders, the deposal of the Ba'ath regime was inseparable from the ideal of freedom of religion. The liberation of Iraq's religious identities had major ramifications for post-war politics and society. In many respects the reclamation of Islam became deeply intertwined with Shi'a liberation from Ba'athism. Shi'a constituted a subjugated majority in Iraq since British colonisation in 1918, and the suppression of Shi'a religious identity had been a major component of Ba'athist rule (Dawisha 2009). With the Ba'ath regime overthrown, Islam became a rallying point for the public articulation of Shi'a identity and discontents both at local and national levels. The most visible announcement of this resurgent identity unfolded in a mass pilgrimage to the holy city of Karbala in late April 2003. The pilgrimage marked the anniversary of the death of the Imam Hussein, perhaps the most beloved prophet in the Shi'a faith, and between one and two million Shi'a travelled, many for days, to Karbala to partake in the ceremony (Cockburn 2007: 89–90). The pilgrimage solidified an important link between Shi'a liberation and Islam; the ability to celebrate one's faith openly became intertwined with the emergence of a Shi'a community released from the shackles of Ba'ath control. The US wanted to characterise liberation in terms of a movement toward democracy; yet Shi'a communities reassembled within a primarily religious narrative. According to Alissa Rubin, 'the administration wanted to tell the story of budding democracy . . . There were some elements of it, but there was also a sort of eruption of long suppressed religious feeling and populist politics' (cited in Hoyt and Palatella 2007: 46).

The outpouring of religiosity did not, however, signify a clearly defined Shi'a religio-political platform (Nasr 2004). Rodger Shanahan explains that Shi'a politics became increasingly crowded in the post-war era and, despite comprising the largest population demographic, Shi'a parties struggled to unite and cement their authority on the national stage during the first year of the occupation (2004: 952). In addition, the main Islamic Shi'a parties were either divisive or fragmented. The Supreme Council for the Islamic Revolution in Iraq (SCIRI), for instance, was closely allied with Iran and therefore distrusted by Shi'a nationalists; while the Da'awa party had fractured into a number of – sometimes antagonistic – sub-divisions throughout the 1990s (Allawi 2007). In short, although Shi'a communities in the main believed that the Shi'a majority should lead the new Iraq, there was no unified vision of what Iraq's future should look like. Instead, what ultimately unified Shi'a political parties was the perceived threat to Islam posed by the US-driven Iraqi constitution.

The creation of a constitution was central to the US conception of Iraqi democracy. In fact, the CPA explicitly stipulated that the possibility of elections was dependent on the adoption of a new constitution. According to Bremer, '[t]he Iraqis don't have a constitution. They need one, and you really can't get to sovereignty without elections, and you can't have elections without a constitution'.[18] In some respects, the US constitutional plans reflected Walzer's conception of the creation of minimal structures. Iraq lacked a robust democratic infrastructure, and many of its existing legal precepts were integral tools of Ba'athist oppression

(Dawisha 2009: 209). Following from the US's own constitutional heritage, and prior experiences in Germany and Japan, the drafting of a constitution appeared to be the next logical step in the creation of Iraqi democracy. In Walzer's terms, the US was attempting to implement their maximalist interpretation of the minimal foundations necessary for Iraqi self-determination. Nonetheless, the US plan was increasingly seen as a concern by the Shi'a religious orthodoxy, and especially by Grand Ayatollah Ali al-Husayni al-Sistani, the highest Shi'a religious authority in Iraq. The story of the US constitutional plan more explicitly highlights the problems implicated in Walzer's conception of the development of minimal structures. Walzer wants to convey a depiction of post-war justice in which building the structures necessary for self-determination is a negotiation between local norms and minimal values. Yet US constitutional ambitions in Iraq illustrate why the maximal articulation of minimal structures can be interpreted as an illegitimate attack on a community's shared life.

As previously outlined in the discussion on the Sadrist movement, the deposal of the Ba'ath regime left a vacuum of authority in Iraq. Shi'a communities soon began to turn toward the mosques and the religious orthodoxy based in the city of Najaf for leadership. To the fore of the resurgence of organised religious authority was Grand Ayatollah al-Sistani. Sistani was an Iranian descended from a long line of Shi'a clerics. An exceptional seminary scholar, Sistani soon became the primary source of Shi'a authority in post-war Iraq (Shadid 2006: 224–226). Sistani's conception of the relationship between religious and political authority, however, greatly differed from the more direct political ambitions of the Sadrists. In contrast, Sistani belonged to the Quietest tradition in Shi'a Islam, which advocates minimal involvement in daily politics. Although Sistani wanted to see Shi'a Iraqis in charge of the political process, he warned his followers not to resist the occupation, advising them to adopt a neutral stance (Nazir 2006: 56). In other words, Sistani called upon the Shi'a majority to bide their time and seek to take control through the election process. Sistani's position in the immediate aftermath of the invasion suggested that the Najaf orthodoxy was happy to accept the occupation providing that it culminated in a Shi'a-led Iraqi government free from US influence. Nevertheless, this did not mean that Sistani and other prominent clerics were willing to accept the radical transformation of Iraqi society that the US envisioned. Allawi explains that the Quietest endorsement of political disengagement is predicated upon the assumption that the political system is already underpinned by Islamic values (2007: 208–210). In a similar manner to Walzer's claim that authentic self-determination is designed to protect minimal rights, Sistani viewed politics as a positive good only if it served to protect Islam. Now we can see the constitutional debate in terms of a contestation of what universal morality means. On the one hand, the US articulated universalism through the Western ideal of liberal secular democracy and, on the other hand, the Shi'a orthodoxy viewed Islam as a universal framework central to any authentic Iraqi state. In Walzer's terms, two maximal interpretations of minimalism came head-to-head in the constitutional debate.

The perceived threat posed by the US constitutional plan stimulated Sistani to engage in more overtly political activities. Sistani responded directly to the US refusal to hold elections prior to the drafting of a constitution by issuing a *fatwa* on 25 June 2003. The *fatwa* explicitly stated that the forces occupying Iraq had no right to draft a constitution or appoint the members of any constitution-drafting body prior to elections (Al-Rahim 2005: 53). More significantly, Sistani's actions suggested that he was willing to undertake a more directly active political role if he felt Islam was at risk. Sistani's political importance was demonstrated in January 2004 when over 100,000 Shi'a took to Baghdad's streets calling for elections and by the beginning of 2004, Sistani had emerged as a key political figure in Iraq and a potential rallying point for Shi'a politics (Jamail 2008: 96). Although the US could not ignore Sistani's political influence, they nevertheless refused to fully accede to his demands. Instead, they hoped to foster a compromise through the creation of a proto-constitution, the Transitional Administrative Law (TAL). The TAL was drafted in early 2004 by the IGC, under the authority of the CPA. Bremer played a key role in the drafting process and the TAL was signed into law on 8 March 2004 (Wheatley 2006: 535). On the surface, the TAL was designed to be a temporary constitution that could be redrafted by an Iraqi parliament once elections had been held – a minimal foundation amenable to local reconstitution. The US argued that the TAL provided the necessary legal infrastructure to hold democratic elections and protected Iraqi self-determination by offering the possibility of a redraft. However, the TAL also symbolised an implicit attempt to solidify the socio-political foundations of the emerging Iraqi state: it was written in English, based on secular values and Islam was described as *one* source of law rather than *the* primary source of law (Allawi 2007: 220–222). Crucially, the TAL contained a clause granting Kurdish Iraqis a veto over any subsequent redraft. Sistani was not appeased and publicly rejected the TAL on the grounds that the Kurdish veto was counter to democratic rule, and that the document was inattentive to Iraq's religious values (Rahimi 2004: 16).

The implementation of the TAL signalled to Sistani and Shi'a Islamic political parties that if they wanted to have any chance of creating an Iraqi society underpinned by Islam they would have to develop a more coherent political platform. The SCIRI had previously attempted to cultivate a collective position through an organisation called The Shi'a House. However, a unified Shi'a political alliance did not emerge until Sistani aligned himself with the group (Allawi 2007: 343). Sistani's involvement in The Shi'a House facilitated the creation of the United Iraqi Alliance (UIA), a broad alliance of Islamic Shi'a political parties. The UIA was dedicated toward the election of a Shi'a-led, Islam-orientated, Iraqi government, and promised the creation of a non-radical stable Iraqi state (Rubin 2005: 70). The final piece of the UIA jigsaw arrived in August 2004 when Sistani brokered a peace settlement between the US and Muqtada al-Sadr, bringing the Sadrist movement firmly under the UIA banner. With the Sadrists on board, the UIA conjoined all the primary Islamic Shi'a political groupings in Iraq in a formidable electoral force backed by Iraq's major Shi'a religious authorities. Nevertheless, the UIA did not mark the cementation of a singular Shi'a politics. Instead, it indicated an alliance built around Sistani's call for Shi'a to unite in defence of Islam. In a post-war society that lacked any established basis for political support, and where questions of who had the right to rule prevailed, alignment with Sistani provided a clear pathway to political legitimacy (Filkins 2009: 246). In contrast, Shi'a parties who refused to ally with the UIA were relegated to the margins of Iraqi politics. Decisively, Sistani issued another *fatwa* on the eve of the January 2005 election proclaiming voting (presumably for the UIA) to be a religious duty for all Shi'a (Jamail 2008: 264). Sistani's endorsement of the UIA ultimately cemented a Shi'a politics in which religion and ethno-sectarian identity were the primary ideological signifiers.

The invasion helped produce an understanding of society in which freedom from Ba'ath repression was directly tied to freedom of religious expression. As religious voices and institutions became the dominant form of post-war authority, Shi'a communities and politics began to rally around Islam as an organisational focal point. In this sense, the political structures that the US sought to develop dovetailed into resurgent Shi'a religious identities, helping to create an expression of politics unified around religious authority and values. The evolution of Shi'a politics in response to religious populism and US constitutional plans highlights a major problem with Walzer's ideal of creating minimal structures. As Walzer acknowledges, minimalism can only be expressed through the idiom of a particular maximal morality (1994: 9). The TAL and wider CPA legislation articulated a distinctly Western conception of society and politics: Western-styled democracy backed by free-market economics and secular liberal values. This reconfiguration of Iraq's socio-political frameworks triggered a defensive response within the Shi'a community. In this respect, attempts to create minimal foundations risk the production of new socio-political contexts, which in turn impact upon how politics and self-determination unfold. In Iraq, US attempts to produce a Western-styled Iraqi democracy solidified Shi'a politics under the dual banners of ethno-sectarian identity and Islam. Intervention thus cannot be justified in defence of self-determination, because it alters social configurations and reconstructs the ideals of self-determination that emerge in post-war societies.

The 2005 election and ethno-sectarian violence

While witnessing the success of the UIA, similar election blocs were formed by Kurdish parties and secular political parties funded by the US (Allawi 2007 391). In contrast, Sunni communities felt completely alienated by the process and responded by boycotting the election, and Sunni candidates who refused to boycott became targets of violent attacks (Dawisha 2009: 254). Owing to fear of violence, the vast majority of candidates contested the election anonymously. As such, the only identifiable features of the election were sectarian identities and agendas (Jamail 2008: 263–264). The election results mirrored the ethno-sectarian divide: from 275 seats, the UIA won 148, the Kurdish bloc 75 and the secular bloc 40 (Allawi 2007: 292). The election further fragmented Iraqi politics in line with ethno-sectarian goals: Sunnis refused to acknowledge the government, the governing Shi'a bloc focused on redrafting the constitution, Kurdish representatives fought for constitutional safeguards to protect autonomy and the US retained a veto on the appointment of any minister they deemed too partisan or divisive. In total, Iraq's first elected parliament represented a number of distinctly ethno-sectarian groupings, coupled with US interests, largely viewing politics as a means to cement a particular image of Iraq. Far from unifying internal fragmentations, the Iraqi political system cast each ethno-sectarian group as distinct, increasingly separatist, electoral cleavages.

The January elections indicated a willingness by Sunnis to shun the political process. The October 2005 referendum on a new Iraqi constitution, however, cemented Sunni disillusionment with Iraq's democratic process. Opposed by Sunni politicians and communities, the constitution enshrined Shi'a conceptions of Islam in Iraqi law and finalised Kurdish autonomy (Cockburn 2007: 196).[19] More importantly, the constitution was ratified despite Sunni populations voting en masse to reject it (Dawisha 2009: 252). In many ways, their inability to block the constitution finalised the realisation of Sunni fears that Shi'a and Kurdish interests would prevail at the expense of those of Sunni communities. Seeing themselves as marginalised from society and politically impotent, Sunnis increasingly began to turn to violence as a means of political empowerment. In the context of their perceived political impotence, violent resistance appeared as one of the only ways in which Sunnis could have any impact on Iraqi society, and the resistance shifted from a broad articulation of Iraqi rejection of the occupation to a Sunni rejection of the Shi'a-led government. This rejection of the democratic process did not solidify into the formation of a unified resistance. Instead, it retained its fragmented and incohesive character. From May to October 2005, for example, 103 different Sunni groups claimed responsibility for attacks (Filkins 2009: 235–238). Nor did it mean that Sunni communities approved the targeting of government forces; for instance, a January 2006 poll found that despite 88 per cent of Sunnis approving attacks on US troops, 77 per cent of Sunni respondents disapproved the targeting of Iraqi security forces.[20]

The January 2005 elections heralded the beginnings of a violent struggle between the Iraqi government and Sunni resistance. Immediately after the election the government began to incorporate Shi'a and Kurdish militias within Iraq's security forces. The main militias incorporated were the Kurdish Peshmerga,[21] the Sadrist Mehdi and the SCIRI's Badr Brigade, a militia group with strong ties to Iran.[22] The Shi'a militias soon began to dominate important sections of Iraq's security forces. For example, US officials estimated that 90 per cent of police in northeast Baghdad were affiliated with the Sadrists by the end of 2005 (Allawi 2007: 423). However, the force that proved most controversial was the Special Police Commandos, a counterinsurgency unit under the authority of the Ministry for the Interior. The Commandos were comprised of 10,000 former Badr Brigade members under the leadership of Bayan Jabr, a former head of the Badr Organisation (Napoleoni 2005: 201). The unit constituted the largest contingent of the new Iraqi Army and was directly charged with tackling resistance groups. For their part, the resistance began to increasingly target state security forces; for example, 616 police officers were killed between January and April 2005 (Cockburn 2007: 192). More importantly, the shift toward Iraqi security forces recontextualised

Shi'a perceptions of resistance violence, which was no longer seen in terms of the rejection of US occupation; it now signified a refusal, by Sunnis, to allow Shi'a to take legitimate control of the country (Allawi 2007: 385).

Resistance violence escalated throughout 2005 and the tipping point came on 22 February 2006 when resistance fighters blew up an important Shi'a shrine in the Sunni town of Samarra. The response by the state security forces, and angry Shi'a communities, was ruthless: in the days following the bombing fifty Sunni mosques were burned to the ground, Sunni prisoners were lynched in Basra and over 1,300 Sunni bodies were discovered in and around Baghdad (Cockburn 2007: 206–207). Over the next few months, Ministry of the Interior forces were accused of acting as sectarian death squads, and reports started to intensify that Commandos would swoop into Sunni areas, killing civilians and taking prisoners (Filkins 2009: 120). In addition, US troops discovered a Commando-operated Baghdad prison holding physically abused Sunni prisoners (Allawi 2007: 422). The governmental response to resistance violence perpetuated an incessant cycle of revenge and retribution, and by autumn 2006, 110 to 130 people were dying every day in Iraq via targeted sectarian attacks (Dawisha 2009: 262). Iraq was suddenly in the grip of a bloody civil war and the country was being redrawn on ethno-sectarian lines. Mixed communities began to dissolve as large numbers of Iraqis migrated to regions where communities shared their ethno-sectarian identity. Sunni and Shi'a populations began to retreat within themselves in the hope that homogeneity would provide peace and security. By the end of 2006, Iraqi society had been radically reconstructed in terms of its ethno-sectarian divisions, and politics was entrenched around these divisions. The ethno-sectarian divide has remained long after the US withdrew from Iraq. The April 2013 elections, for instance, were marred by Sunni claims that voting was postponed in Sunni provinces because of the sectarian motives of the Shi'a government (Arango 2013). In fact, the current war being waged along the Iraqi and Syrian borders by the Islamic State of Iraq and the Levant (ISIL) in the name of the creation of a Sunni state is a direct evolution of the ethno-sectarian tensions that emerged in the aftermath of the US invasion. In this context then, the trenches drawn in response to the invasion and occupation have emerged as the defining feature of Iraqi politics and society in the post-Ba'ath era.

Conclusion

Walzer justifies war on the grounds that it can restore the minimal foundations of self-determination without altering the ability of the community to build a shared maximal life. The evolution of Shi'a politics and Sunni resistance in response to the occupation illustrates the problems with this conception of post-war justice. The ideal of replacing tyrannical regimes and restoring self-determination may seem noble, and even ethically unproblematic; however, the practical implications of this ideal, in Iraq, led to the cementation of ethno-sectarian identities and the reproduction of violence. Key to the failure of US post-war policy is the impossibility of articulating minimal values. Walzer argues that post-war justice is a minimal ideal involving the creation of the universal structures necessary for authentic self-determination. Walzer depicts minimal morality as universally shared and accepted norms – yet minimal ideals, as explained in the previous chapter, are inarticulate and require maximal interpretation. In other words, we never actually see minimalism because it can only be represented through a maximal visage. In this respect, reconstructing the minimal structures of self-determination is not a negotiation between universalism and local particularism. It is a negotiation between particular interpretations of local norms and a particular interpretation of universalism: it is a political battle to define what the community is, who is a member of the community (and who is excluded) and what values underpin communal life. In short, war is a battle to determine the meaning of community.

Walzer begins from the premise that one set of ideals are universal and the other set are particular – his minimal/maximal dichotomy. Nevertheless, the purported minimal ideals are not necessarily recognised as universal because they can only be presented in a maximal vocabulary. As such, minimalism has no role in post-war justice because it never actually appears. Instead, Walzer's ideal of post-war justice is better understood in terms of competing maximal articulations of what a just society means. Iraq provides an example of this form of competition. Conflict escalated because the US, the Shi'a religious orthodoxy and the Sunni resistance all believed that they were articulating universal values: liberal democracy, Islamic precepts, Iraqi nationalism, and so on. Minimalism does not offer a way forward, in this respect, because it does not actually exist. Iraqis did not accept Walzer's purported universal foundations because they interpreted them as an attempt by the US to enforce Western values. Ultimately, the US conception of just resolution helped produce a context in which the Shi'a defence of Islam and Sunni alienation collided in a violent contestation over Iraq's future identity. The chain of events that was set in place by the invasion fundamentally changed the ways in which Iraqi politics and self-determination could take place.

The main point to be taken from this analysis of post-war Iraq is that just resolution cannot be viewed as a defence of self-determination. War and intervention have the potential to radically alter the terrain in which politics and community take place; war risks changing society and the way people interact with their society. War changes the character of the community it seeks to liberate from tyranny. It does not protect self-determination because it changes the nature and identity of the self who engages in politics. Walzer's depiction of just resolution, therefore, cannot justify the call to war; his justification of war begins from the assumption that there is a universal understanding of what a just society looks like, and he justifies war in cases where there is a 'strong possibility' that violence can protect or produce a just society. These twin certainties assure us that we will know when war is the morally right course of action. The case of Iraq, however, calls both of these certainties into question. The transformation of Iraqi politics and society tells us that we do not have control over the types of socio-political formations that war creates. Walzer's certainty that we can control war's outcomes, that there is a strong possibility we will produce just societies through war, is problematic. The US could not control the emergence of Shi'a religious populism or the evolution of Sunni resistance – and attempts to control Iraqi society in this way would have entailed a level of socio-political control that exceeded Ba'ath repression. If US post-war strategy and policies had been better planned, this may have reduced the levels of social antagonism and violence. Nonetheless, better planning and policy execution could not have ensured that Iraqi society remained unaltered by the invasion and occupation. War always risks unintended and unforeseeable consequences. Walzer's twin certainties simply do not work because there is no universally agreed understanding of what a just society entails, and there is no way to ensure that the just outcome is achieved. In this respect, Iraq highlights a conception of just resolution cast in terms of *différance*: Shi'a, Sunnis, Kurds and the US all presented differing interpretations of what Iraqi society should look like. The clashes between these interpretations reproduced a cycle of violence that deferred the emergence of the type of peaceful community Walzer envisions in his justification of war. Iraq illustrates why the noble intentions of a just war cannot guarantee a just resolution. Importantly, we cannot ensure that war will produce justice because we have no way to control how other people respond to the new socio-political contexts that war and violence create.

Notes

1 Lieutenant General Sir Stanley Maude (1917), 'The Proclamation of Baghdad', http://wwi.lib.byu.edu/index.php/The_Proclamation_of_Baghdad

2 Brian Schmidt and Michael C. Williams (2008), 'The Bush Doctrine and the Iraq War: Neoconservatives Versus Realists', *Security Studies*, 17(2), pp.191–220.

3 Project for the New American Century (1998), 'Letter to Bill Clinton on Iraq', http://www.informationclearinghouse.info/article5527.htm

4 For example, see Colin Powell, 'Speech to the UN on Iraq' (2003), http://www.washingtonpost.com/wp-srv/nation/transcripts/powelltext_020503.html

5 For a full transcript of the speech, see George W. Bush, 'Address to the UN General Assembly on Iraq', 12 September 2002, http://edition.cnn.com/2002/US/09/12/bush.transcript/

6 For a full transcript of the speech, see George W. Bush (2003), 'Speech to the American Enterprise Institute', http://teachingamericanhistory.org/library/index.asp?document=663

7 It should be noted that Walzer has subsequently claimed that the US did not live up to its *post bellum* obligations (2012: 44). However, Walzer's critique of US actions in Iraq is concerned with the practical implementation of policy rather than the moral imperatives underpinning the Bush Administration's overarching goals.

8 Jean Bethke Elshtain (2002), 'A Just War?' *Boston Globe*, http://www.boston.com/news/packages/iraq/globe_stories/100602_justwar.htm

9 Most notably, Ahmed Chalabi was one of America's key sources on Iraqi possession of WMDs. See Seymour Hersh, 'Selective Intelligence: Donald Rumsfeld has his own Special Sources. Are they Reliable?', *The New Yorker*, 12 May 2003.

10 Iraq and Iran constitute the two largest Shi'a populations in the world.

11 Quote taken from *Guardian* website: 'Full text: Press conference hosted by US president George Bush, British prime minister Tony Blair, Spanish prime minister Jose Maria Aznar and Portugal's prime minister Jose Durao Barroso, The Azores, March 16, 2003', http://www.guardian.co.uk/world/2003/mar/17/iraq.politics2

12 For a full historical account of military operations in Iraq, see John Keegan (2010), *The Iraq War: The Military Offensive, from Victory in 21 Days to the Insurgent Aftermath* (London: Hutchinson).

13 In addition, CPA Order 17 ensured that US troops were not subject to Iraqi law, meaning that the US Military were not accountable to any authority other than their own government.

14 Infamously, the US Military protected the Iraqi Oil Ministry from looters, reinforcing public perceptions that America's key interest was Iraqi oil resources (Hoyt and Palatella 2007: 31).

15 I will primarily employ the term 'resistance fighter', when possible, rather than 'insurgent'. I adopt this term because the violent resistance began in Iraq before a government was legally constituted, and the aims of resistance groups in Iraq are far more fragmented than a singular focus of overthrowing the government. In this sense, resistance retains a necessary dimension of ambiguity that is lost in the more technical term insurgency.

16 Al-Sadr's father, Grand Ayatollah Mohammed Mohammed Sadiq al-Sadr, had originally been endorsed by the Ba'athists as an Iraqi Shi'a nationalist opposed to Iran and its supposed links to the Shi'a orthodoxy based in Najaf. However, as the Ayatollah's popularity grew, he began to vocally oppose the Ba'ath regime, culminating in his assassination alongside two of his sons (Cockburn 2007: 92).

17 Sadr employed this ideal as a means to undermine the political ambitions of returning exiles who fled Iraq under Ba'athist rule to escape punishment (Napoleoni 2005: 136).

18 This quote is taken from a PBS Newshour Interview with Paul Bremer on 24 September 2003, http://www.pbs.org/newshour/bb/middle_east/july-dec03/bremer_9-24.html

19 In turn, the Kurdish regional authority drafted their own constitution that superseded the Iraqi one (Dawisha 2009: 264).

20 Figures taken from Worldpublicopinion.org (31 January 2006), 'Poll of Iraqis: Public Wants Timetable for US Withdrawal, but Thinks US Plans Permanent Bases in Iraq', http://www.worldpublicopinion.org/pipa/articles/brmiddleeastnafricara/165.php

21 The Peshmerga became the regional military and police force in the Kurdish autonomous region.

22 The Badr Brigade initially began as a unit of Shi'a Iraqis trained by the Iranian Military during the Iran–Iraq War, but became an exclusively Iraqi outfit during the post-Gulf War Shi'a uprising (Katzman 2008: 170).

3 Derrida and ethics

Introduction

The first chapter characterised Walzer's conception of the world as auto-affective, an ontological system underpinned by the self-determining subject. Walzer's morality, at all points, serves as a means to protect internal coherency. Be it in human beings or the political community, self-determination is the thread that binds Walzer's image of the world together. Morality, in Walzer's terms, is a regulatory system that allows self-determining political communities to exist and flourish. The defence of self-determination, however, should not be confused with isolation and withdrawal from alterity; Walzer is not advocating islands of maximal life fully detached from their outside. Rather, he positions the existence of the self-determining subject as a structural necessity for any meaningful engagement with alterity. The world starts with internal coherency and radiates out toward external engagement. Nevertheless, Walzer conceives alterity as something that should never be directly implicated in the production of the subject.[1] Importantly, Walzer's ontology suggests that the outside carries a lingering threat: the outside threatens to intervene and forcibly corrupt the natural process of self-determination. In other words, the outside is desirable so long as it remains clearly separated from the coherent inside. It is only by fixing alterity in its proper place, outside the borders, that we can ensure justice. In this sense, the idea of just war is captured in a singular motif: a war that produces or protects a self-determining communal subject. In turn, the rhetoric that underpinned Bush Administration justifications of the Iraq War was couched in this ideal of securing Iraqi self-determination. The concept of regime change, as such, was primarily framed in terms of producing a government that authentically represented the Iraqi people. Yet the previous chapter highlighted the problematic incorporation of this view of ethics within the US plan for post-war Iraq. In attempting to cultivate internal coherency, the US drew attention to Iraq's ethno-sectarian divisions, leading to an increasingly fractured population, distrustful of difference.

This book presents a model of ethics that challenges the ontological primacy of the self that is evident in Walzer's justification of war. I call this alternative conception of ethical responsibility *ethics as response*. The term *response* is crucial to the understanding of ethical responsibility discussed in this chapter because it emphasises the relationality implicated in responsibility. In this respect, ethics as response highlights how the self is produced in relation to other people.

This understanding of ethics does not begin with the ideal of an internally coherent subject; rather, it begins by unpacking the constitutive and supplementary relationship between the self and other. The central argument in this chapter thus contends that subjectivity begins within the grip of ethical relationships. While response alludes to the question of responsibility (what is my duty to others?), what is more important to the present argument is the play of movement indicated by the term. The theme of movement is essential to the following argument as it suggests an image of ethics that begins in flux, rather than from within a stable self, and a model of responsibility that cannot be satisfied: an acknowledgement that ethics is always in transition and never fully resolved or resolvable. In certain respects, this understanding of ethical responsibility is a form of Derridean ethics; yet Derrida's ethics is itself articulated as a response – being, for example, a response to the work of Plato, Immanuel Kant, Georg Wilhelm Friedrich Hegel, Kierkegaard, Martin Heidegger, Emmanuel Levinas, Nancy and others to whom his arguments are indebted. Derridean ethics is, therefore, another chain of supplements: it does not begin with Derrida – it is produced through Derrida's responses to other authors. In this sense, the term response describes a model of ethics that is never the sole property of the singular author; an already fractured subject resounds within the concept of ethics as response.

The primary aim of the following discussion is to illustrate how understanding ethics in terms of response challenges the model of justice expounded throughout Walzer's work. Key here is the assertion that by questioning the metaphysical assumption that ethics begins in the self-determining subject, we can open the possibility of producing alternative conceptions of what it means to act ethically. Expanding upon the Derridean theme of the supplement introduced in the first chapter, the subsequent discussion articulates an understanding of ethical responsibility that begins within the constitutive relationship between self and other. Above all, this chapter seeks to demonstrate the impossibility of satisfying or completing our responsibilities toward other people, thereby moving us toward a new understanding of ethical responsibility that advocates sustained engagements with the consequences of our actions. Ethics as response does not attempt to close the question of ethics with a general set of moral rules and laws; instead, it demands a sustained engagement with the questions of ethics, action and justice.

Ethics as first philosophy

Bulley contends that the question of ethics is, fundamentally, the question of how we relate to otherness (2009: 3). However, this understanding of ethics, as Bulley acknowledges, relies upon an already determined conception of self and other; thus, the ethical relationship presupposes a categorically separable multiplicity of subjects. For instance, in Walzer's model we start with a coherent self who is a member of a political community and is, therefore, capable of recognising its ethical duties to both members and strangers. In this context and in order to understand the Derridean response to the question of ethics, it is important to start with the question of subjectivity or, more precisely, how subjects who can act ethically are produced. The auto-affective tradition, to which Walzer is indebted, conceives subjectivity as a mode of immanent self-determination. For example, the Platonic subject has an innate capacity for knowledge and reason from which ethical duty arises (Plato 1987: 260–262), and the Cartesian subject engages with the outside world through self-recognition of its own consciousness (Descartes 1998). In this respect, the auto-affective tradition contends that the self can engage with the outside world because it is assured of its own subjectivity. Self-consciousness is positioned as the originary moment of subjectivity – and human beings are subjects because they are capable of self-reflective thought.

Auto-affection describes an ontology in which the world is offered to the self by the self; the world becomes meaningful to the self by virtue of its own volition. As Derrida contends, the auto-affective model presupposes a self that is conscious and certain of its relationship to itself, i.e. pure-auto-affection (1997: 97–98). In this sense, auto-affection constitutes a mode of existence in which the self is the foundation of all ontological experience. We start with the self-conscious subject who reaches out to the world to draw in objects and render them meaningful to itself. Other people are one type of object that the self draws in and, through this movement, the self develops meaningful relationships. The ideal of ethical responsibility that follows auto-affective thinking takes the form of reciprocal substitution: the duty the self has toward others is premised upon the understanding that the other is a subject *like* the self. Ethics starts within the self and reaches out toward those who resemble the self; it is an ethics derived from familial recognition. This is the crux of Walzer's understanding of ethics: moral duties defined by the meanings, values and heritage we share with others. This is why Walzer argues that we have wide-ranging ethical duties to members of our own community in contrast to the minimal duties we owe to strangers. Nevertheless, by looking at the disintegration of Iraqi society, we begin to understand the problems with this ideal. Defining ethics in terms of familial relatedness risks solidifying alterity as a source of danger, which in turn, structures society along the fault lines of difference. In Iraq, this has driven ethno-sectarian violence, entrenchment and the withdrawal from engagement with alterity.

Looking toward Heidegger's conception of being can help us bridge the gap between auto-affective understandings of subjectivity and the Derridean critique.[2] Heidegger moves away from auto-affection by presenting us with a conception of being that begins in relation to its outside. Yet *Dasein* (Heideggerian being) still maintains the primacy of self-relation: '*Dasein* is a form of self-relation which is systematically connected to others of the same kind, others of different kinds' (Hodge 2001: 2). In certain respects, Heidegger breaks with consciousness as the origin of subjectivity while retaining self-relation as the nucleus of existence. *Dasein* remains a self-relation, but it is a self-relation that is part of a wider system of external relations. As such, *Dasein* repositions the self-relation as always already in the midst of its relation to its outside, thereby destabilising the certainty that self-consciousness is the origin of meaning. Heidegger contends that *Dasein* constitutes a mode of existence in which the self takes care of itself through the objects that it encounters (1996: 322–325). In Levinas's terms,

Dasein is a form of 'corporeal existence', because the subject lives through what is other than itself (2008: 164). In the Heideggerian model, the relational aspect of existence is rooted in the ways in which subjects relate to the objects from which they live; for example, the use of wood, a hammer, nails, straw, and so on to build a home that shelters the subject from the elements. Thus, meaning is produced within this relationship between the self and the objects through which it lives. However, this model of existence is predicated upon self-recognition of finitude and mortality. In the absence of this recognition, there would be no impetus for subjects to render objects meaningful as a means to protect their existence. In this sense, consciousness is possible precisely because existence is at stake. In Heidegger's terms, *Dasein* is an existence in which being is thrown toward its own end (1996: 233). Heidegger wants to stress that objects do not hold inherent transcendental meanings; instead, meaning is produced through this relationship between objects and finite beings. Objects, in this respect, are meaningful by virtue of the self's relation to its own death, giving rise to the Heideggerian understanding of subjectivity as singularity-unto-death. Derrida expands on this theme by explaining that being is singular and irreplaceable solely by virtue of its own death: 'Everyone must assume their own death, that is to say, the one thing in the world that no one else can *either give or take*' (2008: 45, original italics). This is not to say that people cannot be saved, or save others, from a specific risk of death; for example, that we could not prevent another person from drowning. Rather, singularity-unto-death means that death can be delayed but it cannot be definitively escaped. Death is imminent and unavoidable in the case of every single person. In Heideggerian thought, conscious existence is an expression of the drive to delay the moment of death – the subject engages with the world as a means to preserve its own existence. In Levinas's words 'to be conscious is to have time' (2008: 165–166).

Time brings us to a pivotal point in Heidegger's conception of subjectivity. For Heidegger, temporality grounds the possibility of subjective existence (1996: 335). Temporality allows the self to render the world meaningful in relation to its own finitude – therefore, *to be* is to exist within a structure of finite time: '*To exist is to be "temporalized"*' (Levinas 1996: 13, original italics). In order to understand the foundation of the subject then, we need to understand how being is temporalised. According to Heidegger, being is self-temporalising, thrown by itself into the truth of its own existence (1998: 252). In other words, the self recognises its own finitude, and this allows the self to render the world temporal, finite and therefore meaningful. Although being is thrown into existence in relation to other beings and objects, the Heideggerian model starts with the self-revelation of finitude. Following from this, the relationship with other beings remains antecedent to self-consciousness and this, in turn, produces a philosophy that privileges the self-relation over ethical relationships: 'this thinking is not ethics in the first instance because it is ontology' (Heidegger 1998: 271). In this way, self-temporalisation becomes foundational to Heidegger's ontology, and meaningful existence radiates from this particular self-relation. For Heidegger, the subject is born, through self-temporalisation, into a world that offers the tools of care necessary to delay the moment of death. The objects offered to the subject are properly understood as objectives of self-possession, objects through which the self can live. This birth of subjectivity, in the Heideggerian context, is synonymous with the advent of language: what Heidegger terms as 'the clearing-concealing advent of being itself' (1998: 249). Language is the name Heidegger designates for the self's relation with temporality that allows the subject to relate itself to the external world of objects. This understanding characterises Heidegger's image of language as the home in which the subject dwells (1998: 239). Language, as being's home, signifies a self-conscious subject with the capacity to draw objects into its consciousness and render them meaningful in relation to its own finitude. What is important for Heidegger is the idea that language represents the possibility of self-possession. Language provides a home into which the subject can draw objects of care; in turn, these objects help prolong the subject's existence. Crucially, in Heidegger's thought, the self's possession of objects is never in question. However, language also represents the originary possibility of ethical relationships: a subject who can recognise and possess objects of care is presented with the possibility of offering care to others. In Derrida's words, language is the ultimate homeland through which we can reach out to the other in a hospitable, ethical gesture (2000: 89).[3] Language, in this sense, creates the possibility of a self that can possess objects of care which can subsequently be distributed to others. Language, as home, is the possibility of relating ourselves to other people – the possibility of ethics (Bulley 2009: 64–65).

Heidegger's ontology is removed from the spontaneous, auto-affective, emergence of the subject. Yet it simultaneously retains the self-relation as the originary meaning of subjectivity. It is therefore unsurprising that Derrida's challenge to Heidegger is articulated through an alternative understanding of language and communication. What is primarily at stake in this contestation is the way in which being becomes temporalised. While Heidegger positions language as the advent of the self-temporalised subject, Derrida conceives language as prior to the event of temporalisation. For Derrida – and Levinas – language does not signify the subject's self-relation to finitude. Instead, the commitment of an approach toward language (that does not originate in the self) underpins the possibility of subjectivity. Levinas describes language in terms of the proximity of one to the other and the commitment of an approach (1999: 5) – in other words, language is defined as the possibility of a relationship between subjects. For Derrida, the self and the other exist in a relation that pre-exists their subjectivities – the self-relation with finitude – and, therefore, precedes their thematic separation into categories of self and other. This re-presents the assembly of the conscious subject in terms of an a priori relation to alterity: the movement toward subjectivity is always already an approach toward the other. Derrida reminds us that this does not cast the other as the origin of subjectivity. Rather, the relation is itself the origin and as such, remains a non-origin. In Derrida's terms, the relation is already in movement and therefore cannot be the starting point (1997: 19). In contrast to Walzer, Derrida does not conceive the self-determining subject as the origin of meaningful existence. Instead, subjectivity is made possible through the spatial relation symbolised by an approach toward the other. This movement toward alterity is also an approach toward language and communication; thus, Derrida describes the relation between subjectivity and language in terms of spacing rather than Heideggerian temporality:

> *Spacing* (notice that this word speaks of the articulation of space and time, the becoming-space of time and the becoming-time of space) is always the unperceived, the non-present, and the nonconscious . . . It marks *the dead time* within the presence of the living present, within the general form of all presence.
>
> (Derrida 1997: 68, original italics)

Spacing indicates an originary relationship between being and its outside prior to consciousness and subjectivity. The becoming-subject of being is predicated upon the constitutive relation with what is other than being; the self becomes a self because it is already related to its outside. Yet, as will be explained, this pre-originary relation is retained, as a trace, within every act of consciousness. In this way, the Heideggerian conception of language and self-temporalisation is replaced by a relational movement toward alterity – what Levinas describes as an anarchic trauma that opens consciousness from the outside (1999: 123).

To this point we have articulated the relational foreword to subjectivity in rather abstract terms. Therefore, in order to ground the relation in a more concrete light, we must revisit the concept of finitude, the Heideggerian locus of subjectivity. For Heidegger, consciousness of death signifies the becoming-time of being that allows subjects to render the world meaningful. However, the self's experience of its own death is a conscious experience only to the extent that it nullifies self-consciousness. In Nancy's words, it marks the becoming other of consciousness (1991: 33). Hence, finitude cannot be conceived as a self-relation through which conscious subjectivity is assembled. Instead, finitude is properly understood as a relationship with the outside from the beginning. As Levinas explains, the subject can only witness the death of other people and this exposure to death is intimately tied to ethical responsibility (2008: 179). Derrida builds upon this motif by arguing that consciousness of death starts with the recognition of mortal others, and this repositions the self-relation with finitude as a relationship with alterity: 'The relation with the other and the relation with death are one and the same opening' (1997: 187). Cast in this way, finitude describes a mode of existence predicated upon exposure: collective exposure to finitude creates the possibility of subjective existence. Nancy terms this exposure as *comperance*: 'finite existence exposed to finite existence, co-appearing before it and with it' (1991: xl). What is important is the acknowledgement that a singular subject abstracted from alterity could not become aware of its own mortality. If there were no others the subject could not possibly be conscious of its own mortality because its only experience of death would be its own, which would nullify self-consciousness. Exposure to other people's mortality, therefore, allows the subject to understand itself as finite and opens consciousness toward temporality. It is through exposure, as Levinas maintains, that the self is provoked as an irreplaceable singularity-unto-death from the outside (1999: 105–106). As such, the Heideggerian understanding of ontology – the relation between objects and finite being – is subverted. The ethical relationship with other people precedes the self-relation with death.

The mutual exposure to finitude that opens subjectivity correlates to the Derridean understanding of language and the home. For Derrida, language signifies the constitutive relationship between meaning, finitude, alterity and subjectivity: 'it is the principle of death and of difference in the becoming of being' (1997: 25). The relationship between the becoming-subject of being and ethical relationships is explained through the concept of substitution. Here, substitution should not be conceived in terms of the auto-affective model of familial reciprocity in which the subject recognises the image of the self in the other and acts ethically on the basis of this recognition. Instead, substitution marks the moment in which the self finds itself at home, in language, through the other. It is because subjects are exposed to each other that communication and language become possible and, therefore, that the categories of self and other appear. In this sense, the conscious subject exists by virtue of its exposure to others: the host enters their home (becomes conscious) through the guest (Derrida 2000: 125). The home, which as we recall signifies the possibility of ethics, is possible because of the relation to alterity. What is paramount to the Derridean conception of substitution is the idea that language and meaning do not emanate from the subject's free will. For Derrida (2002a), substitution constitutes a decision that is made about us before we even have the possibility of deciding: the decision of the other in me. What Derrida means by this phrase is that the subject does not consciously decide to embrace language and meaning. Being is thrust into language by virtue of its exposure to the outside prior to its assembly into conscious subjectivity. Levinas characterises this model of substitution in terms of being as hostage: 'It provokes this responsibility against my will, that is, substituting me for the other as a hostage' (1999: 11). Although Levinas portrays substitution as responsibility in opposition to free will, substitution is better understood as prior to will. Derrida describes the opening of subjectivity in terms of passive openness to alterity (1997: 240). The possibility of being-at-home-with-oneself, of language and subjectivity, is affirmed by the image of a home that is already open to the coming of the other. Language, meaning and subjectivity are possible because being is always already open to alterity. In Derrida's words,

> Without a trace retaining the other as other in the same, no difference would do its work and no meaning would appear. It is not the question of a constituted difference here, but rather, before all determination of the content, of the *pure* movement which produces difference. *The (pure) trace is difference.*
>
> (Derrida 1997: 62, original italics)

The possibility of a meaningful existence is, in other words, underpinned by a trace of alterity retained within the self. Fundamentally, substitution indicates alterity rooted at the heart of subjectivity, a self whose roots are twofold. Levinas characterises this formation of subjectivity as a response prior to the question

(1999: 25). What Levinas means by this is that the other has a hold over the self prior to the formation of conscious subjectivity. The presence of the other within the self is simultaneously a demand for a response. Following from this, the opening question of being is not a question of ontology– 'who am I?' – but instead, the opening questions are of ethics: 'who is this other?' and 'how will I respond?' Yet these questions are not temporally separated and their simultaneity indicates consciousness that begins already in question.

In contrast to Heideggerian being, which offers itself meaning in the mode of self-preservation and self-possession, subjectivity formed in relation to alterity is delivered into a world where meaning is contested, a priori, in the form of a question. Levinas describes this calling into question of the self, which is simultaneously the assembly of the self, as ethics (2008: 43). Ethical exposure is not equivalent to the Heideggerian exposure of being to objects. Instead, exposure to ethics signifies subjection to the vulnerability of other people, which throws back the trembling image of the self's own vulnerability to the self: substitution 'is the subject's subjectivity, or its subjection to everything, its susceptibility, its vulnerability' (Levinas 1999: 14). This model of subjectivity is cast in terms of subjection to vulnerability through exposure to alterity, a pulsation of being-nothingness-becoming that beats under the regime of exposition (Nancy 1991: 88–89). Importantly, the understanding of objects as a means through which being can live is contested at its roots, because objects are also a means through which the other can live. The vulnerability of the self is produced in relation to vulnerable others who are equally dependent on objects as a means of existence. Thus, other people call into question my joyous possession of the world by exposing me to their vulnerability, their susceptibility to trauma and to death (Levinas 2008: 75–76). The subject does not only recognise its own finitude, it also recognises that self-possession is implicated in other people's vulnerability and death. The subject recognises that how it engages with the world impacts upon others. Yet this calling into question is precisely what opens the possibility of self-possession. Exposure to vulnerability opens being to finitude and the ability to render objects meaningful as tools of self-care. In this respect, to exist as a finite temporalised subject means to be exposed to a world of questioning – that is, the inwardness of subjectivity presupposes external exposure to the questioning of the other; the Cartesian subject knows himself because he is exposed (Nancy 1991: 31). Following from this, ethical responsibility does not begin in a movement from self-possession to giving in which self-possession is assured. Subjectivity begins in a world in which possession is already contested by the presence of the other. *Dasein* is being originally with others; and concern for others, not concern for objects, is the constitutive determination of subjectivity (Nancy 1991: 103). It is exposure to the other's vulnerability that opens the possibility of self-care and, as such, the self begins as a response to alterity.

Existence starts from a position in which the self is already in response to alterity; to be a subject is to be exposed to my responsibilities to others. This understanding recasts responsibility in the form of a general responsiveness to the questions posed by others, thereby supplanting the conception of responsibility that starts with self-identification and familial reciprocity:

It is only in approaching the Other that I attend to myself . . . in discourse I expose myself to the questioning of the Other, and this urgency of the response – acuteness of the present – engenders me for responsibility; as responsible I am brought to my final reality.

(Levinas 2008: 178)

Levinas points toward subjectivity in which the self is pre-ordained as responsible for the other, a self that is always implicated in others' care and survival. Yet to be responsible is not tantamount to an inherent benevolence toward others. In fact, exposed to its own vulnerability, the self is likely to view other people as threatening because they represent the most direct contestation of self-possession and, therefore, survival. In this respect, responsibility does not correspond to an innate passive goodness. Ethical relationships are not analogous to a pre-destined moral duty, and the call for a response remains a call that can be rejected. The self can refuse to respond; it can neglect other people in order to preserve itself. Instead, ethics as response contends that the subject is responsible for what has not begun in them: ethical relationships. In short, to be responsible is to be in a position capable of responding to others without having chosen to be in this position.

Community as the possibility of justice

The reconceptualisation of subjectivity as a response to alterity provides the impetus for a rejection of Walzer's justification of war as a defence of community. The Derridean understanding of a self that exists in response to others challenges the auto-affective ideal of community central to Walzer's justification of war. In the first chapter I outlined and critiqued Walzer's depiction of self-determining communities. As I explained, Walzer justifies war as a defence of self-determination because he believes that community is the singular irreplaceable element of existence. While particular individuals are dispensable, a particular collection of people living life according to their shared meanings, values and social distributions can never be replaced (2005: 49). Community is irreplaceable in Walzer's ontology because it establishes the foundations necessary for the creation of a maximal world and, therefore, meaningful existence. Following from this, ethical responsibility is entirely predicated upon community: neither maximal nor minimal morality can exist without the foundation of communal belonging. Community is the absolute structure of Walzer's ontology and ethics – meaning, politics, ethical relationships and morality are possible because human beings exist within separated political communities. Importantly, for this discussion, Walzer conceives communities as self-determining from their beginning. Yet Walzer also stresses that communities are not naturally occurring formations – they are socially constructed. As such, what is natural in Walzer's model is the predisposition of people to form communities as a means to protect self-determination. For Walzer, people want to live life according to their own values – and community provides the sole means to secure this mode of existence. As previously discussed, community, in Walzer's ontology, is founded upon the commitment of a set of people to live according to their own beliefs and values. Community starts from the commitment toward common meanings and is oriented toward the realisation of these meanings. In this way, Walzer's community is both auto-affective and immanent: the communal subject comes into being through its commitment to become itself and is realised through its self-solidification. The foundations and underlying structures of Walzer's ontology are underpinned by this movement and ethics is conceived as derivative to self-determination: ethical responsibility is engendered and satisfied in the formation and preservation of the self-determined communal subject. Therefore, maximal and minimal moralities are conceived as safeguards designed to protect self-determination.

By recasting the subject in terms of a constitutive interplay between self and other, the Derridean conception of ontology provides a challenge to Walzer's model. Although we have touched on this contestation, it will be illuminating to further clarify the Derridean critique of auto-affection, and to explain how this is linked to the idea of ethics as response. Recalling the Derridean understanding of communal foundation, community begins in the midst of a performative and constitutive violence that solidifies an inside of community by demarcating the outside. For Derrida, the foundation of community cannot derive from pre-existing common values because the performative inauguration creates the very basis of these values: it creates the coherent inside in opposition to the threatening outside. In this sense, self-determination is founded through the expulsion of alterity, and this expulsion constructs the self. The violent opening of community is, therefore, simultaneously a closure: the exclusion and suppression of what is designated other than community – that is, community can only become what *it is* through the exclusion of what *it is not*. In one sense, Walzer is correct in suggesting that the right of closure is intimately related to the possibility of ethics. Without the separated subject there are no knowable others and, therefore, no possibility of actively responding. However, in the Derridean understanding of ethics, the foundation of community mirrors the formation of subjectivity: the communal subject comes into being as a response to alterity. This is a departure from Walzer's auto-affective ideal because it highlights why self-determination is impossible without the pre-existing relation to alterity. Community is not self-determining in the first instance; it is determined by the constitutive interplay that forms the categories of inside and outside community – members and strangers. Nonetheless, in the absence of a subject clearly delineated from others, there is no possibility of meaningful existence. Hence, the existence of the communal subject is underpinned by separation and the sovereign power to exclude others. Derrida reminds us that no community – or subject – can identify itself without exclusion; the outside must be demarcated in order to render the inside meaningful (2002b: 57). What is important in Derrida's argument is that we are faced with a relational, rather than auto-affective, origin of community. Community does not spring from self-determination; instead, it is constituted through a relationship that demarcates the boundaries between inside and outside. The opening of self in relation to others creates a structure in which the self is in question, vulnerable and exposed. The response of the self to vulnerability is to seek refuge by closing itself off from others – substitution necessarily denotes separation. In this respect, the promise of the founding act of exclusion is that the communal subject can remain what *it is* without fear of external interference.

Yet the constitutive role of alterity in the foundation of community suggests that alterity is, perhaps, something retained within community without the possibility of definitive exclusion; that is, if community is impossible without the relation to alterity, can community still operate if alterity is completely excluded? Walzer's conception of justice is particularly illustrative in this context. For Walzer, a just community remains grounded on the possibility of internal disagreement and dispute; self-determination remains in tension between differing voices. In its foundation, the communal subject cuts itself off from alterity in the name of self-determination, only to find that alterity is itself a pre-condition of justice. Authentic self-determination necessitates different ideas of what the common life should entail. In this sense, Walzer does not view self-determination as a drive toward complete unity. Instead, he wants to emphasise the defence of commonality: members are not fully unified, but they share certain meanings and values that distinguish them from the strangers outside. In Walzer's argument, the primary thing that members have in common is language. Language facilitates the emergence of common meanings, values, and social understandings that combine to form a shared structural foundation through which it is possible to negotiate a common life. The alterity implicated within the formation of community is, in this way, domesticated through the ideal of a common language. Although the community has different ideas of what maximal life should entail, disputes are understood through a common framework.

Walzer presents an image of community tied to a particular understanding of communication: the communication of a shared life that is not altered through its transmission from one member to another. In contrast, Walzer argues that communication between strangers lacks an articulate common language and is only united, on special occasions, through the invisible assemblage of minimal morality. Walzer's understanding of the way in which a community communicates is synonymous with what Derrida describes as the logocentric ideal of full and original speech, 'a unanimous people assembled in the self-presence of its speech' (1997: 134). Again, what is necessary for Walzer's ideal of community to make sense is a conception of language as self-presence: maximal language is fully present to members of a community and binds them together, despite their differences. It is through the self-identical commonality of language that members can distinguish between internal others (members), who complement self-determination, and external others (strangers), who threaten to supplant self-determination. Shared language differentiates the internal contestations of common life, which make justice possible, from the external contestations, which threaten to destroy community. Language demarks the coherent core of subjectivity, which allows Walzer to domesticate certain forms of alterity, while simultaneously excluding alterity that is not amenable to the coherent inside. The defence of self-determination is thus, at bottom, the defence of a particular understanding of the relationship between language and alterity. Justice, for Walzer, means keeping unassimilable alterity separate from the maximal life. To act morally, therefore, is to keep undomesticated alterity, by force of law and war, in its proper place – outside a community of members unified through common language.

The conception of alterity outlined by Walzer is directly related to Derrida's understanding of the supplement discussed in the first chapter. In this case, alterity is a dangerous supplement that promises to protect justice while, simultaneously, threatening it. In Walzer's argument, alterity (contestations to the common life) is a necessary component of any just community and is positive, if it can be domesticated through common language, and negative, if it cannot. Common language, defined in terms of self-present common meanings and values, signifies the possibility of taming the supplement and rendering it as an uncomplicated addition to community. In short, the ideal of common language allows Walzer to present internal contestations of maximal morality as a positive communal good – while justifying war as a defence against external contestations. The Derridean understanding of language, however, retains the supplement in all its ambiguity: communication without the possibility of homogenisation. In this respect, Derrida's understanding of communication suggests that language can never be decisively domesticated and, therefore, retains a trace of the undomesticated and the external. Derrida (1988) unpacks his understanding of communication through the concept of *iterability*. *Iter*, meaning 'other' in Sanskrit, points toward the otherness of communication. Derrida argues that communication is inseparable from alterity because every act of communication cuts itself off from self-consciousness as the ultimate authority of the transported meaning (1988: 7–8). Derrida explains that communication necessitates that the communicated message remains readable, recitable and repeatable – in other words, communication presupposes a future relationship with another person who will be called upon to interpret the meaning of what is being communicated.[4] Language, in this way, remains in relation to alterity by risking-to-mean something other than what is intended by the sender (Derrida 2002a: 242).

In Walzer's terms, if we want to share language between members of a community, the common language must risk reinterpretation in order to remain articulate. Walzer's ideal of communication presupposes a homogeneous space of inscription; a space of fully unified meaning and understanding between members, or what Derrida terms 'a homogeneous element through which the unity and wholeness of meaning will not be affected in its essence' (1988: 3). Walzer's understanding of shared language thus allows members of a community to disagree and dissent, while maintaining the internal coherence of the subjects discussed. Nevertheless, Derrida contends that language itself can only open toward communication within the structure of iterability, which signifies the alterity necessitated in communication. To function as communication, language needs to be cut off from its originary presence (the sender) and meaning needs to be separated from the intention that guides its production (Derrida 1988: 5). The separation of meaning from its origin in communication implies that alterity is a necessary component in the possibility of all communicative movements. Rather than fixing common meaning, language can only function through the process of risking-to-mean something other than it intended. Without this risk of corruption no communication would be possible, because the risk is essential for the movement of meaning from sender to the receiver. Communication necessitates the loss of full control over the meaning and, in this respect, it opens language to interpretation and contamination. As Derrida contends, the desired homogeneity of communication 'ruins itself and contaminates itself; it becomes a spectre of itself' (2002a: 277). In this way, Walzer's certainty of shared meanings and common values is undermined by the interpretative structure of the communicative gesture. Membership is founded on the promise of a common language, yet common language is always already contaminated by the risk of alterity.

Iterability suggests that the way in which language is communicated between members is structurally indistinguishable from the way it functions in relation to non-members. This is not to underplay the role of social and cultural reference points and communal memory. It is simply to state that self-present commonality, the complete domestication of language, is an inaccessible ideal. Meanings may, in certain cases, be more common between members, but all communication nevertheless exposes language to the risk of untameable alterity; communication exposes language to reinterpretation. As such, language – between members and strangers alike – functions by virtue of its relation to alterity: sharing language necessitates the risk of meaning becoming other-than-intended. Recasting alterity as the possibility of shared language challenges Walzer's conception of community and his justification of war. Because it is no longer possible to distinguish between domesticated alterity that complements self-determination and the *foreign* alterity that threatens internal coherence, it is no longer possible to justify war in defence of an illusionary ideal. Community shares a language that is always contaminated by undomesticated alterity and functions in a manner that is structurally indistinguishable from the language shared between strangers. That is to say, community is not characterised by self-determination; it operates through mutual contamination. Community, as the communication of commonality, is possible only if it remains in relation to undomesticated alterity. Derrida (2002a) explains mutual contamination under the theme of immunity and auto-immunity. The communal subject views alterity as threatening and, therefore, strives to exclude alterity from its boundaries to protect itself. Yet alterity marks the possibility of the subject's existence, so the exclusion of alterity also risks the destruction of the self. In this way, the self strives to protect itself through closure to alterity (immunity), but definitive closure would entail stasis and the death of the self (auto-immunity). In Derrida's words, '[i]t conducts a terrible war against that which protects it only by threatening it, according to this double and contradictory structure: immunity and auto-immunity' (2002a: 82). Auto-immunity describes Walzer's conception of community in which alterity is excluded in the name of self-preservation: it is excluded in the name of the pure realisation of self-presence. However, the realisation of fulfilled self-presence signifies the immobilisation and death of the self (Derrida 1988: 128–129). In contemporary Iraq we can see the logics of auto-immunity played out on a grand scale. Each facet of Iraqi society believes that its identity and values are being assailed by threatening others. This fear of alterity has, in turn, fuelled the violent defence of communal identity: each side justifies its own violence in defence of self-preservation. This attempt to defend the communal self, nevertheless, has been massively destructive and continues to annihilate the possibility of a stable and peaceful Iraq. The drive to preserve the self, in this way, threatens the self's existence.

Subjectivity, as we have explained, is predicated upon the relation with alterity that calls the self into being by calling it into question. Translated into Walzer's terms, a just communal subject implies a community in which meanings and distributions are contested. Abstracted from the questioning of the other, the self has no requirement to exist in the active sense: nothing is contested, so no decisive actions are required. This is why Derrida argues that community is above all else the community of the question (2002b: 356). Walzer's primary question is that of justice: how can we justly defend our communities in a morally acceptable way? And his response suggests that justice is preserved if we keep alterity within its appropriate place – outside community's borders. Yet Derrida (2002a: 295) reminds us that the absolute exclusion of alterity is tantamount to the extermination of the demand for justice – the elimination of a communicative call for a response:

> The violence of injustice has begun when all members of a community do not share, through and through, the same idiom. Since in all rigour, this ideal situation is never possible . . . The injustice, which supposes all the others, supposes that the other, the victim of the injustice of language, if we may say so, is capable of a language in general.
>
> (Derrida 2002a: 246)

Derrida's critique of communal unity has important implications for the possibility of justice. The model of justice advocated by Walzer justifies violence as a defence of self-determination, yet this justification is predicated upon the illusionary ideal of common language. Community claims to justly exclude others on the basis of a common language that is fully shared between members. However, if language can be shared, it must be iterable. In this respect, Walzer's common language is always in the process of becoming other than what it intends to be. The injustice of common language is that it excludes others on the basis of alterity, while simultaneously retaining alterity as the basis of its very functionality. Walzer's ideal of a common language, then, both presupposes and violently excludes alterity. The form of justice Walzer presents is more adequately understood in terms of community as the possibility of immortality. It is just for individuals to risk their lives and the lives of others in the defence of community because the sacrifice allows the essence of their values and common life to live on through the community they have created. In this sense, community exists as a surrogate vessel for collective subjectivity: some part of the finite subject can live on through its work. In turn, Walzer argues that the collective work (the community) is more valuable (irreplaceable) than that of the individuals who construct it. Walzer's justification of war suggests that the continuation of life lived according to a particular set of values and social distributions is more important than life itself. In his argument, the sacrifice of life is justified in the name of the continuation of a particular way of living.[5] We find in this model a re-articulation of the redemptive sacrifice of Christ's crucifixion: everlasting life purchased through cruel death (Asad 2007: 85). In this way, the sacrifice of life is erected as a monument to the anticipated eternal life of the community.

The communal subject's drive toward immortality, however, is simultaneously a death drive – not simply in terms of the individual deaths justified in defence of the collective, but in terms of the auto-immune risk inherent in the pursuit of absolute self-preservation. The completion of a fully self-determining subject necessitates the annihilation of justice: Walzer's ideal of a contested maximal life. A community that becomes absolutely unified loses its capacity for being-in-common – its capacity for ethical relationships (Nancy 1991: xxxix). If community becomes a singular unity there can be no differentiation between the social spheres and, therefore, no possibility of justice. This is not the system that Walzer wants to valorise. Rather, he wishes to differentiate between familial others and foreign others through the idea of common language. Yet, as we have already discussed, this differentiation cannot be sustained in a rigorous manner. The irreplaceability of community is a mythical narrative – common values and meanings are always embroiled in the process of communication; they are always in the process of being re-written, re-interpreted and replaced. As such, the self-determining essence that Walzer wants to protect through the sacrifice of war is always already in transition toward alterity; it is never fully determined by the self because the self is itself determined, in part, by alterity. Walzer, therefore, justifies the sacrifice of individual lives in the defence of an impossible ideal. Ethics as response inverts this by presenting justice as an experience of the impossible. Justice resembles a perpetually unsatisfied call that finds no rest, a call that is always in transmission/transmutation toward the other and the other others.

Ethical action as sacrifice

Thus far this chapter has primarily focused upon demonstrating the necessity of alterity in the assemblage of subjectivity, and explaining why this challenges Walzer's justification of war as a defence of self-determination. However, this is not to imply that the concept of self does not have a role to play in Derrida's understanding of ethics. In this respect, the Derridean understanding of responsibility attempts to recontextualise Heidegger's concept of singularity-unto-death: the singularity of the self is re-imagined as the active component of responsibility; the ability of the subject to respond to others.

Derrida's understanding of ethical action emphasises the relationship between death and responsibility. Central to this idea of responsibility is the correlation between finitude and singularity: 'Death is very much that which nobody else can undergo in my place. My irreplaceability is therefore, conferred, delivered, "given", one can say, by death' (2008: 42). While this relationship is the foundation of Heidegger's ontology, Derrida highlights the ethical dimension: 'only a mortal can be responsible' (ibid.). He turns the Heideggerian image of death inside out, by suggesting that singularity is intimately tied to the subject's capacity for ethical action. In this sense, finitude is no longer seen as the means through which being uncovers its final truth. Instead, death becomes the means through which ethical responsibility is produced as a mode of being. Derrida explains that the subject exists alongside death and alongside mortal others; and this reflects the alterity within the self that sets the self in motion. Consciousness of death is, in this way, re-inscribed as the possibility of being responsible to other people, because it opens the possibility of responding to the needs of others as a means to delay the moment of their death. Yet the possibility of acting ethically is underpinned by the non-negatable structure of death. In Derrida's words, death is 'the one thing in the world that no one else can *either give or take*: therein resides freedom and responsibility' (2008: 45, original italics). What Derrida means by this is that death, as the non-transferable singular property of the subject, opens the possibility of ethical action. Specifically, exposure to death creates a structure in which the subject is responsible – free to respond to, or ignore, the needs of others. The self's exposure to its own death through others opens the possibility of acting for or against the other. Singularity through substitution opens toward ethical action.

Derrida describes ethical action in terms of *se donner la mort*, the gift of death. He asks,

> How does one give *oneself* death? How does one give it to oneself in the sense that putting oneself to death means dying while assuming responsibility for one's own death, committing suicide but also sacrificing oneself for another, *dying for the other*, thus perhaps giving one's life by giving oneself death, accepting the gift of death, such as Socrates, Christ, and others did in so many different ways.
>
> (Derrida 2008: 12, original italics)

Derrida's argument should not be understood in the sense that dying for the other is the only form of ethical action. Instead, what he wants to emphasise is that the spectre of death creates a mode of existence in which life is at risk. Thus, it is only when life is at risk that other people require my assistance; it is only in a situation where the other's existence is at stake that we are called to respond. Here we see more clearly the relationship between responsibility, gift and death. The inevitability of death puts life at risk, creating the possibility of giving to others in an ethical gesture. Derrida reminds us that death symbolises the possibility of giving and taking while, simultaneously, exempting itself from this structure (2008: 45). In this way, death institutes the need to act in order to sustain life – the life of the self or the life of others. The spectre of mortality, therefore, creates the conditions through which the subject can give or take objects and put them to use, to protect their own life or the lives of others. Yet death itself remains outside the system it inaugurates. The subject's own death can never be given to another or taken by another, and every single subject must undergo a singular death which is theirs and theirs alone.

Se donner la mort, Derrida suggests, can also be discussed in terms of hospitality. Hospitality presupposes a home that can be offered to a guest and the singularity of the home defines the possibility of a subject capable of welcoming another. This image signifies the understanding of ethical action expounded in this chapter: the subject at home-with-itself through its singular property, its death, can give to the other through the hospitable/ethical gesture. To welcome the other into the home is to give to the other in the form of ethical action. 'The singularity of the home,' Bulley argues, 'should not be given up because, while it can be a violent "closedness", it is also the very condition of openness, of hospitality and of the door' (2009: 64). Hospitality, therefore, describes the way in which the separated subject is open to the other; and this marks the possibility of giving to the other – the possibility of ethical action. Hospitality, nevertheless, highlights the tension between ethical action and ethical responsibility. Ethical responsibility is understood as absolute openness to alterity. Yet the hospitable gesture is only possible through closure, exclusion and violence. It is only possible to offer hospitality when the self has the ability to close the home and choose between others. In Derrida's words, there is '[n]o hospitality, the classic sense, without sovereignty of one self over one's home, but since there is also no hospitality without finitude, sovereignty can only be exercised by filtering, choosing, and thus by excluding and doing violence' (2000: 55). What is imperative to the Derridean understanding of hospitality is the singularity of the hospitable gesture. Singularity is important in two senses: the singularity implicated in the possibility of the sovereign home; and the singularity of the decision to welcome specific others.

Derrida explains this dynamic by drawing our attention to the two laws of hospitality. The first law is that of unconditional hospitality – complete openness – while the second law signifies conditional hospitality, granted to specific others on specific terms (2000: 75–77). Derrida maintains that neither of these laws can function in and of themselves. The unconditional law destroys the possibility of the home and sovereignty because it corresponds to passivity and indifference; that is, if the home is unconditionally open it is no longer possible to offer hospitality, because it erases the sovereign subject who can offer a welcome. On the other hand, the conditional law must be oriented and engendered toward unconditional, impossible, hospitality: the conditions are meaningful only in relation to absolute openness. For Derrida, the unconditional law is a law without duty, a law without law; while the conditional law is a plurality of laws that corrupt their proposed foundation. The laws are 'both contradictory, antinomic, *and* inseparable. They both imply and exclude each other, simultaneously' (2000: 81, original italics). It is for this reason that Derrida calls for us to understand the ethical movement in terms of a conjoined hospitality and hostility – *hostipitality*: openness to specific others through the exclusion of other others (2000: 45). This is of course commensurate to describing an ethical responsibility that opens unconditionally and passively to alterity, but which can only be directed, as action, in relation to specific identifiable others. Responsibility, in this respect, can only be enacted through exclusion: the generality of ethical responsibility brushes against the grain of the singularity of ethical action.

Hostipitality describes an understanding of ethics that encapsulates the divided seal of subjectivity and the singularity of the self. On one hand, ethical responsibility opens toward the general call for a response: the subject is born in response to others. On the other hand, however, ethical action can only be undertaken in the specific and the singular: the subject must choose between potential responses. The twin pillars of Derridean ethics, general responsibility and singular action, build upon Kierkegaard's reading of the biblical tale of 'The Binding of Isaac'.[6] In Kierkegaard's reading, Abraham is commanded by God (the absolute) to sacrifice his only legitimate male heir, Isaac, for a reason that God keeps secret from Abraham. Kierkegaard contends that Abraham's absolute responsibility to God requires him to renounce his ethical responsibility to humanity.[7] The act is rendered even more horrendous because Abraham accedes to this unthinkable sacrifice without knowing God's reason. Derrida argues that Abraham is simultaneously the most responsible, because he maintains his undivided duty to God without hope of reward and without knowing why the sacrifice is demanded, and the most irresponsible, because he has utterly denounced his ethical obligation to humanity in the name of his absolute bond (2008: 73). In this sense, the moral, and indeed morality, of Kierkegaard's reading is that an individual cannot be absolutely responsible without sacrificing their responsibility to what Kierkegaard terms the ethical.[8] For Kierkegaard, to be responsible in the singular (absolute) sense, one must sacrifice responsibility to the generality (the ethical). This interplay between the ethical and the absolute marks the tension between general and singular responsibility in Derridean ethics. Derrida maintains that because the other is absolutely singular, my responsibility to every single other is also absolute. The phrase *Tout autre est tout autre* [every other (one) is every (bit) other] characterises the Derridean incorporation of Kierkegaard's understanding of ethics: 'If every human is wholly other . . . then one can no longer distinguish between a claimed generality of ethics that would need to be sacrificed in ethics, and the faith that turns toward God alone, as wholly other' (Derrida 2008: 84). In other words, my openness to ethical responsibility is an unconditional openness to alterity and, as such, my responsibility to the generality of others is unlimited. Nevertheless, I can only act in the singular sense, for example, to intervene or respond on behalf of this individual or these individuals. This, as Derrida emphasises, creates a structure in which ethical responsibility is inseparable from irresponsibility:

> There are also others, an infinite number of them, the innumerable generality of others to whom I should be bound by the same responsibility, a general and universal responsibility (what Kierkegaard calls the ethical order). I cannot respond to the call, the request, the obligation, or even the love of another without sacrificing the other other, the other others.
>
> (Derrida 2008: 68–69)

In this way, the singularity of the self is retained through the singularity of action. However, singularity is placed in tension with multiple, innumerable, calls for response. Therefore, I cannot respond to a singular other or particular others without sacrificing my ability to respond to every other in similar or differential positions of peril at the same instant. Ethical action, in this respect, entails decisions that symbolise a necessary dimension of irresponsibility – a dimension of sacrifice.

To conceive ethical action as sacrifice is to implicate the unethical in the ethical movement. Above all, ethics described in terms of sacrifice precludes satisfaction. Derrida argues that sacrifice suggests that ethical obligations can only be affirmed through the debasement of responsibility; to respond to other people in the name of ethics simultaneously necessitates the sacrifice of ethics (2008: 69). In line with the dual laws of *hostipitality*, simultaneity is paramount. It is not a case of the singular action supplanting the ethical generality, allowing responsibility to be satisfied in a singular instance. Instead, the general is retained and recognised at the moment it is transgressed through action (Derrida 2008: 66–67). For Kierkegaard, Abraham must retain his love for Isaac at the moment he commits to murder him, because the ethical must retain all its value to make the sacrifice a sacrifice (2006: 65). General responsibility hence engenders and orients the singular action which broaches and breaches the law of its commencement. People are thus called to respond in a general sense; and this general responsiveness is the foundation of conscious subjectivity and human action. The act of responding, however, necessitates a choosing between different people and different actions. To respond, therefore, means to sacrifice one's ability to respond to every other, in the same instant; and this sacrifice is founded by the ideal of general responsibility. Nevertheless, refraining from ethical action does not resolve or allay the transgression, and ethical responsibility cannot be satisfied in passivity or inaction. Although the subject is called to responsibility passively, this is merely one component of ethics. Levinas describes responsibility through the metaphor of breathing. Breathing symbolises the way in which the subject involuntarily draws the outside in as a means of sustaining its own existence – a passive exposure to the outside. Nevertheless, the call to responsibility that passively penetrates and forms the conscious subject cannot be ignored in the passive sense. Just as one cannot passively refrain from breathing, ethical responsibility can only be rejected in an active sense and non-action remains a decision and, therefore, a sacrifice: the conscious refusal to respond to any other (Levinas 1999: 180–182). In this respect, sacrifice should not be lamented because it is the only possible mode of ethical action (Levinas 2008: 149). Responsibility, my ability to respond to others, propels me into existence, which is always the space of absolute sacrifice. To exist is to live in a space in which we must choose between different others and different responses.

War is perhaps the most evocative articulation of the space of ethical sacrifice because, as Jan Patočka contends, the distance between subjectivity and mortality is most visible (1996: 129–130). In Derrida's words: 'War is a further experience of the gift of death (I put my enemy to death and I give my own life as sacrifice by "dying for my country")' (2008: 19). The relationship between ethics, sacrifice and war brings us to an important juncture in this discussion. What is particularly important is the image of the soldier sacrificing their life as a loving gift in the defence of their country. Sacrifice in the name of the state is central to Walzer's justification of war. This symbol of sacrifice, however, also reminds us that sacrifice is always directed toward something other than the self: it is a sacrifice *in the name of*. Levinas (2008) explains that sacrifice signifies the way in which death takes on meaning through responsibility. Dying for one's country, for instance, becomes understood in terms of heroism. Derrida expands on this theme, arguing that *Se Donner la Mort* also means to interpret and give signification to death (2008: 12). As such, war exemplifies the relationship between ethics and sacrifice: the individual's sacrifice is oriented toward the advancement of some greater good that lies outside the self. This represents the core argument presented throughout Walzer's theory of war: the sacrifice of individuals is necessary in the defence of community. Walzer firmly asserts that war is a crime because it forces men and women to risk their lives in the name of their rights (2006a: 51). So, fighting a just war means that we must sacrifice individual lives in the name of communal life. Walzer's depiction of just war indicates an understanding of ethical action, in which Derrida contends that ethics and sacrifice are conjoined in the name of duty (Derrida 2008: 67). In Walzer's terms, we must risk the absolute rights of some members in the defence of the irreplaceable shared life they have built together. In this way, he describes a mode of ethical action in which acting in accordance with the moral law entails violence toward others. For example, in the context of Iraq, as Bulley suggests, invading means both protecting and attacking Iraqis, sacrificing some Iraqis and US soldiers in the defence of other Iraqis (2009: 48). Walzer thus acknowledges that acting morally in war entails a dimension of sacrifice.

Where Walzer's reasoning departs from the Derridean understating of ethical action is in terms of justification. At bottom, Walzer believes that the signification of the sacrificial gesture can justify the consequences. That is to say, the perceived goodness of the sacrifice can render it morally permissible: specific acts of violence are morally desirable because they are undertaken in the name of a greater good. The JWT, in some respects, presents certain sacrifices of life as wholly moral and permissible, albeit lamentable, provided they match specific criteria. As Talal Asad argues: 'The genealogy of humanist sensibility joins ruthlessness to compassion and proposes that brutal killing can at once be the vilest evil and the greatest good' (2007: 86). Walzer provides us with an understanding of ethical responsibility in which our absolute responsibility to the community justifies the killing of certain individuals in specific contexts. His justification rests on the grounds of a law of salvation: acting violently will save the community. However, Walzer can only make this justification by starting with the self-contained communal subject that must be protected from the outside. Ethical responsibility, for Walzer, is pre-conditioned by a commitment to the self-determining communal subject prior to any call to respond to others. In this sense, his justification of sacrifice and violence against others is auto-affective in design. The sacrifice of life is justified in the name of community because the sustained existence of the community is a necessary pre-condition for the possibility of ethics itself. This means that it is ethically responsible to sacrifice lives in the name of community, because without community, no ethical responsibility or justice would be possible. Again, ethical responsibility starting from the self justifies the violent defence of the self against threatening others.

On the other hand, the Derridean conception of responsibility refuses the ideal of sacrifice in the name of self-determination:

> And I can never justify this sacrifice . . . whether I want to or not, I will never be able to justify the fact that I prefer to sacrifice any one (any other) to the other . . . what binds me to singularities, to this one or that one, male or female, rather to that one or this one remains finally unjustifiable (this is Abraham's hyperethical sacrifice), as unjustifiable as the infinite sacrifice I make at each moment.
>
> (Derrida 2008: 71)

By starting from general responsiveness to others, Derridean ethics refuses to justify sacrifice on the grounds endorsed by Walzer. Instead, the subject opens to an unlimited call for response in which justice is an infinitely unsatisfied right of the other (Derrida 2002a: 257). The subject who acts ethically in a specific instance can only do so by failing in their duty to respond to all the other calls for justice. The sacrifice of life in the name of the community retains the simultaneity of its ethical and unethical dimensions: to violently defend a set of people remains unjustifiable in a definitive sense. Ethics as response, in this way, highlights an understanding of responsibility without the relief or satisfaction derived from justification.

Undecidability as justice for the other

Walzer's argument is that justice is tantamount to the defence of a particular ideal of self-determination, and war is justified in cases where a community's capacity for self-determination is adjudged to be under threat. In this respect, justice is guaranteed if undomesticated alterity is kept, in its proper place, outside a community of members. In Derridean ethics, community is an experience of *hostipitality*, a welcoming of certain others that creates a home by virtue of the exclusion of other others. While Derrida contests Walzer's justification of war, he does not endorse a retreat to absolute pacifism and the abandonment of all closure. This is because the pacifist opening of community to all alterity without resistance does not resolve the question of responsibility. Passive openness to alterity may destroy a particular closure, a particular communal formation, but it does not destroy the system of closure itself – which is directly tied to the possibility of the subject's very existence. Instead, the pacifist opening risks new closures and new injustices. As Derrida's (2002a) discussion on revolutionary violence suggests, the absolute refusal to defend the home against aggression risks the violent formation of a new state or social grouping. In this sense, refusing to defend the home against threatening others potentially risks the creation of socio-political transformations that present new and unforeseen adverse consequences. Following from this, the purpose of Derridean ethics is not to seek definitive closure or openness; rather, it is, as Bulley advocates, a search for better closures (2009: 84). Derridean ethics is thus a questioning of the justifications presented for the preservation or creation of specific ethical and political closures. The questions posed by Derridean ethics are, therefore, how do we decide between different closures? How do we decided between different responses? How do we decide between different sacrifices? How do I act ethically in the knowledge that my actions betray the responsibility I wish to uphold? These questions highlight the trauma of ethical action cut off from the telos of fulfilment – what Derrida (2002b) terms the ordeal of undecidability.

Undecidability is intimately tied to the Derridean understanding of 'decision'. Derrida claims that decision represents the nucleus of the singular subject's capacity to respond to others, and marks a choosing between potential actions that relates singularity to responsibility: '. . . retrenched in one's own singularity at the moment of decision. Just as no one can die in my place, no one can make a decision, what we call "a decision", in my place' (2008: 60). In this sense, the decision to respond or not to respond (to whom to respond and how to respond) describes the way in which the subject is related to its ethical responsibilities. The subject's singular decision to respond in a certain way in the name of specific others describes the structure of ethical action. Here, freedom and responsibility combine in an active sense: the subject is always already called to respond to others but is free to decide if and how it responds. In turn, Derrida argues that negotiation describes the process through which decision unfolds. For him, negotiation signifies the trauma of ethical action in which people can respond to a specific other or others only by neglecting their duty to all others; the irresponsibility lodged within all ethical action. Derrida (2002b) argues that ethical responsibility propels the subject into a situation where they are faced with decisions between multiple actions, all of which potentially risk negative effects. The ethical tension implied in negotiation re-articulates the hyper-ethical sacrifice in which ethical action necessitates a negation of ethical responsibility. Derrida contends that negotiation implies a choosing between incompatible imperatives (2002b: 13).[9] For example, the ethical duty to save Iraqis from Ba'athist oppression could only be enacted by placing the lives of Iraqis at risk; the imperative to protect is coupled with the risk of sacrifice.[10] In this respect, competing imperatives necessitate a response, yet the response can only materialise into action by betraying one or more imperative. As such, negotiation describes a mediated response: the generality of responsibility mediated by the singularity of action. Derrida's conception of negotiation is a departure from conventional models of ethics. The conventional response to the ethical trauma of choosing between incompatible imperatives is to resolve the conflict via the application of moral rules and laws; for example, some variant of categorical imperative, or utilitarian calculation, in the name of a greater good, or Walzer's War Convention. Conventions, then, claim to negate ethical sacrifice by presenting us with a clear framework for morally justified action. Derrida rejects the conventional model of response because it decontextualises decisions. 'There is no general law for negotiation,' Derrida maintains, because '[n]egotiation is different at every moment, from one context to the next' (2002b: 17). In other words, conventional models claim to solve a singular ethical question via the application of a general law, irrespective of the specific context of the case. As I will demonstrate in the following chapters, general rules abstracted from specific contexts deconstruct themselves through their application. US justifications of siege warfare, for example, failed to take account of the specific contexts encountered during the 2004 sieges of Fallujah, or the tangible implications derived from this failure.

In addition to the failure of general laws to take context into account, Derrida points to a larger problem with the application of rules: namely that general rules purport to solve ethical problems with technical knowledge. Derrida maintains that rule-based morality satisfies itself in a form of technics incompatible with ethically active subjectivity. He argues that the deployment of technical knowledge nullifies responsibility by transforming decision into a mechanic relationship of cause and effect (Derrida 2002b: 231). What Derrida wants to stress is that the resolution of ethical tension through the application of knowledge is equivalent to the negation of the ethical relationship itself. The subject that acts in accordance with rules is no longer related to its ethical responsibility; it is not responding to the needs of others. Instead, it is responsible solely to the diligent application of the general rule. For instance, in terms of the categorical imperative, if I were to satisfy responsibility through the application of the general rule, 'I would act, Kant would say, *in conformity* with duty but not *through* duty or *out of respect* for the law' (Derrida 2002a: 245, original italics). In this respect, acting on the basis of a general rule places the subject in a relationship of obedience to the law, rather than in an ethical relationship with those who call for a response. In addition, justification via the application of rules cuts responsibility off at the very moment it is put into action: as soon as the subject has acted in accordance with the moral law, its responsibility has ended regardless of the actual consequences. As such, by justifying action in relation to rules, we distance ourselves from the consequences these actions inflict upon other people. The resolution of ethics through the deployment of codified rules, therefore, contains the seeds of its own dissolution, because it maintains ethical action only by erasing the link between the subject and those affected by its actions. The subject that follows a convention acts in accordance with the general rules, but not in response to specific others in a specific context; the duty is to the rule, not to the other in peril. In Derridean terms, it is a technical application of knowledge and not a responsible decision (2008: 26).

Ultimately, Derrida rejects conventional resolutions of ethical questions because they assume that technical knowledge can provide ethical certainty. Conventional rules assume that we know what the right thing to do is in a given circumstance, and that we are certain of the effects our actions will produce. In contrast, Derrida wants to draw our attention to the dimension of uncertainty implicit to every singular instance of ethical action. He describes the absence of rules and certainty as the suffering of deconstructive justice (2002a: 231). Despite this suffering, uncertainty is not viewed in the negative sense of a barrier to ethical action. Rather, as Levinas suggests, uncertainty propels the subject toward sustained ethical engagement (1999: 20). Derrida expands on this image of uncertainty as a necessary component of ethical action by relating it to the phenomenality of trembling:

> . . . [w]e tremble in the strange repetition that ties an irrefutable past (a shock has been felt, some trauma has already affected us) to a future that cannot be anticipated . . . Even if one thinks one knows what is going to happen, the new instant, the arriving of that arrival remains untouched, still inaccessible, in fact unlivable . . . I tremble before what exceeds my seeing and knowing although it concerns the innermost parts of me, right down to the soul, down to the bone, as we say.
>
> (Derrida 2008: 55)

In his discussion on trembling Derrida links uncertainty to responsibility and action. Being, cast into subjectivity through responsibility, trembles on the precipice of decision because of the uncertainty (the unanticipated future) implicated in action. The subject trembles before what exceeds knowledge. We tremble because we do not know (and cannot know), definitively, how our actions will affect other people. In this way, ethical action is a risk not only because it requires us to choose between others, but perhaps more importantly, because we can never know with absolute certainty that our actions will deliver their intended results. As Paul Ricoeur states, '[j]udgement means that we "shall be judged" on what we have done to persons, even without knowing it . . . That is what remains *astonishing*. For we do not know when we influence persons' (2007: 109 original italics).

What is crucial is the understanding that at the moment of decision a dimension of non-knowledge is lodged within the ethical action: when we decide to act we can never ensure that our actions will achieve exactly what we want them to achieve. This is both the positive condition of ethical action and its negative limit. If I act in full knowledge of the effects of my actions, there is no decision and no ethical relationship; there is just the mechanical application of the rule and obedience to the law. However, if I act without full certainty, I must risk the unintentional and unethical effects my action may induce. The uncertainty implicated in every possible action recalls the alterity implicated within ethical responsibility. In a similar vein to the concept of iterability, ethical action is always in the process of becoming other than its guiding intention through its unforeseeable impacts. Alterity engenders ethical responsibility in the subject and ethics rejoins alterity through the uncertainty of ethical action. Derrida describes the uncertainty in the moment of decision as the undecidable, and maintains that undecidability introduces the incalculable at the heart of ethical responsibility (2002a: 100). Undecidability provides another example of the logic of the supplement introduced in the first chapter. Again, the supplement signifies an irresolvable ambiguity that sets the movement in motion while simultaneously denying the possibility of fulfilment. Without the undecidable there is no possibility of ethical action, yet to act within the frame of undecidability means to risk unethical and unforeseeable consequences. Following from this, Derrida recasts undecidability as the positive impossibility of justice:

> The undecidable is not merely the oscillation of the tension between two decisions. Undecidable – this is the experience of that which, though foreign and

heterogeneous to the order of the calculable and the rule, must nonetheless – it is of *duty* one must speak – deliver itself over to the impossible decision while taking account of the laws and rules. A decision that would not go through the test and ordeal of the undecidable would not be a free decision; it would only be a programmable application or the continuous unfolding of a calculable process. It might perhaps be legal; it would not be just.

(Derrida 2002a: 252, original italics)

This understanding of the undecidable is particularly important as it redefines the relationship between knowledge and uncertainty. The ordeal of the undecidable decision does not imply that we must cut ourselves off from knowledge and calculation. Instead, it describes how an incalculable, unknowable, aspect of action is essential to the possibility of justice. As Derrida suggests, we must strive to know as much as possible and yet, uncertainty remains and must remain (2009: 54).

Derrida (2002a) argues that the relationship between knowledge and uncertainty in ethical action marks an instance of aporia, a non-path that illustrates why justice is an experience of the impossible. Therefore, if we respond to the call of the other – respond to the demand of justice – we must risk injustice. As with sacrifice, iterability, *hostipitality* and auto-immunity, undecidability is another law of positive contamination: there can be no decision in the absence of the undecidable, yet the presence of the undecidable forecloses the accomplishment of justice. Uncertainty, in this way, both sets the possibility of ethical action in motion and blocks the fulfilment of justice. In turn, Derrida maintains that uncertainty cannot be alleviated by future knowledge. It is not possible to resolve the aporia through future knowledge because the future in which such knowledge can be secured has already been determined by the prior decision (Derrida 2009: 56); in short, the decision alters the future and the future knowledge that would resolve the aporia. In this sense, undecidables function as resistance to the closure of the question of justice (Derrida 1981b: 43). As Levinas contends, '[i]t is not I who resists the system, as Kierkegaard thought; it is the other' (2008: 40). The fulfilment of justice is barred through the undecidable, the element of alterity inherent in every ethical action that signifies the becoming other of action through its unforeseeable impacts. Nonetheless, the becoming other of ethical action does not exonerate the subject of its singular responsibility. Ethical responsibility symbolises the decision of the other in the self, but the self cannot abandon ethical action to other people (Derrida 2002a: 56). Instead, ethical responsibility – cut off from the possibility of fulfilment – signals the need for more sustained ethical engagements. As Levinas maintains, it describes a metaphysical desire for ethical action that deepens rather than satisfies responsibility (2008: 34). Ethical action deepens, rather than satisfies, responsibility because we are obligated to follow the consequences of our actions. Conventional models of ethics cut responsibility off at the precise moment the subject has acted in good faith and in accordance with the moral rule; the subject is satisfied that it has done the right thing once it has followed the letter of the law. Derridean ethics refuses to cut responsibility off at the moment of action. In contrast, the undecidable sustains subjects' responsibility through the action's movement toward the other and the other others. The subject must maintain a portion of responsibility for the new conditions created by its actions and how these conditions affect, or are affected by, others. It resembles a categorical imperative in which the ends/means relationship is reconfigured: the ends are de-limited perpetually, deferring the justification of means – a categorical imperative written in terms of *différance*. In this respect, ethics as response demands that we not only follow the response to our actions but also follow the responses to the response. Ultimately, we are responsible for alterity: we are responsible for the ways in which other people respond to the conditions our actions help create. We are called to remain invested in an unlimited chain of actions and consequences, retaining limited responsibility for the infinite eruption of multiplicity in one singular act (Nancy 1991: 102). The other resists the closure of ethics by reminding us that the act is never completed, never finalised. The act remains in perpetual motion so long as there are others that can be affected. This, as will be discussed in the conclusion to this book, has important implications in regard to current Western attitudes to Iraq – in particular, the belief that non-Iraqis have no further role to play and should just let the population get on with the task of rebuilding the country.

The purpose of describing ethical responsibility in terms of response is to emphasise the relationality between the subject, action and alterity. Deconstructive ethics works through what Derrida calls a modality of the 'perhaps' (2002b: 344): *perhaps* my actions have not achieved their aims, *perhaps* I have made the situation worse, and so on. It is this trembling of the perhaps that keeps the question of justice alive. However, the decision to act also risks the possibility of closure. Deconstructive ethics must risk becoming an iterable set of rules and guidelines; it risks becoming a convention. In Derrida's words, '[f]or a deconstructive operation, *possibility* would rather be the danger, the danger of becoming an available set of rule-governed procedures, methods, accessible approaches' (2002a: 264, original italics). To resist this possibility, Derridean ethics advocates an unwavering commitment to account for the specific context in which the call for decision may arise (1988: 136). This is why any decision that portends the possibility of justice must reinvent itself in each case (Derrida 2002a: 251). Each instant of ethical action must follow through a new ordeal of undecidability. Every single decision, ethical action and deconstruction must remain a singular event; every decision must remain heterogeneously other. Yet no one can sustain the question of justice, of responsibility, of sacrifice, of hospitality, or of subjectivity. These questions call for others to respond, to resist closure. Ethics as response calls upon the other, through the medium of ethical action, to keep the demand for justice infinitely open and unsatisfied.

Conclusion

This chapter has sought to map out an understanding of ethics that emphasises the relational interplay and movement between self and other, and responsibility and action. The central argument is that understanding ethics as response offers a better way forward for thinking about war. This idea of responsibility opposes models of ethics that proceed from auto-affection. Derridean ethics points toward a relational/ethical foundation of subjectivity that is grounded upon a constitutive movement toward alterity. More precisely, it describes a mode of existence in which the subject is always already related to itself by virtue of its relation to others. In other words, the subject is responsive in a general sense prior to the formation of the categories of self and other. To be a subject is to exist as a response to others. However, the generality of responsibility is complicated by the singularity of action: the unlimited call for response can only be enacted in terms of singular responses to specific others. This brings us to an understanding of ethics enacted through sacrifice: ethics becomes action only through the sacrifice of the general obligation to respond to all others. This movement broaches and breaches ethical responsibility by way of a supplementary logic that marks the impossibility of fulfilling justice. Supplementary logic, as such, correlates to the positive condition and negative limit of ethical responsibility: to act in the name of ethical responsibility means to sacrifice the fulfilment of justice. Undecidability describes the way in which irresolvable uncertainties rooted in the moment of ethical action preclude the possibility of justice. This absolute uncertainty propels ethical action into a future epoch in which the subject must retain limited responsibility for the ways in which its actions intentionally and unintentionally, directly and indirectly, affect others. Responsibility is engendered and maintained through alterity: the subject borne in response to others must follow the consequences of, and responses to, its actions without the possibility of definitive satisfaction or closure. The subject begins as response and continues in response to others.[11]

The following two chapters attempt to situate the arguments presented thus far in a more specific and grounded context: that of the Iraq war. In this sense, the subsequent chapters expand upon the themes introduced here by illustrating the ways in which Derridean thought can allow us to better understand how ethics are produced and enacted in war. Key to this exposition are the ways in which ethical relationships are engendered and transformed through action, and why this challenges Walzer's conception of justice. The next chapter focuses upon the justification of intentional killing in war, and Walzer's rationalisation and moralisation of the killing of combatants. This chapter demonstrates why the killing of combatants signifies unjustifiable sacrifice rather than a moral imperative, thereby displacing Walzer's foundation of the moral rules of war. The fifth chapter focuses upon the principle of 'double effect', Walzer's justification of the unintentional killing of civilians in war. The primary purpose of this chapter is to highlight why the unintentional effects of war cannot be dismissed in terms of accident and error and are, therefore, a complicit component of what it means to justify violence.

Notes

1 Walzer accepts that the outside can influence the inside, but only to the extent that the inside chooses to incorporate an external idea by virtue of its own free will.
2 Particularly important is Heidegger's (1996) *Being and Time*.
3 This image is intimately related to Walzer's ontology in which being at home in a political community is a foundational necessity for the possibility of morality.
4 Derrida maintains that this structure remains in cases where the sender and the receiver are the same person. For example, in the case of a person writing a shopping list for themselves, the production of the list is intended to compensate for a future absence of memory (Derrida 1988: 49).
5 However, as will be discussed in the next chapter, Walzer does not view the loss of life in war as a sacrifice of ethical responsibility.
6 See Søren Kierkegaard, *Fear and Trembling* (2006), trans. Sylvia Walsh, editors C. Stephen Evans and Sylvia Walsh (Cambridge: Cambridge University Press). The full tale of the binding of Isaac is contained within Genesis 22: 1–19.
7 This is underscored by the fact that Isaac is claimed to represent the future promise of Abraham's people. In this sense, Isaac symbolises the future promise of humanity (Kierkegaard 2006: 14).
8 By the ethical, Kierkegaard means any proposed universal morality. Specifically, in *Fear and Trembling*, he is responding to the Hegelian understanding of morality.
9 Derrida's argument is related, in some respects, to Walzer's conception of 'Dirty Hands' (1973) and Supreme Emergency (2006a, 2005). Walzer contends that in certain instances we are forced to choose between two morally problematic courses of action. However, Walzer believes that such instances are rare (moral emergencies) and that we can resolve the issue, in part, through calculation: 'I can argue that I studied the case as closely as I was able, took the best advice I could find, sought out available alternatives. And if all this is true, and my perception of evil and immanent danger not hysterical or self-serving, then surely I must wager. There is no other option; the risk otherwise is too great' (2006a: 260).
10 The decision not to intervene places other Iraqis at risk. Negotiation characterises an implicit choosing between different responses, which impact different others in different ways.
11 This movement is no less evident in instances in which the subject chooses to act unethically or even in denial that any ethical responsibility exists. The negation of ethical responsibility is, nonetheless, marked by an acknowledgement that responsibility has, to a certain extent, been negated. To deny or renounce that any responsibility exists maintains the relation to the others potentially affected by one's actions. As Levinas maintains, one cannot absolutely or definitively negate the ethical relation; the negation retains its meaning as negation, whether this is recognised by the self or an other (2008: 198).

4 Non-combatant immunity and the sacrifice of rights

Introduction

The previous chapter outlined an understanding of ethics in which ethical action entails a simultaneous broaching and breaching of responsibility: one cannot act ethically without sacrificing responsibility. Importantly, this broaching and breaching of ethics precludes the possibility of justice, or definitive justification. Ethical responsibility, in this respect, is an experience of the impossibility of justice. The last chapter also provided a critique of systems of morality that aim to resolve questions of ethical responsibility through the deployment of codified rules. The critique stated that rule-based models of morality divert ethical responsibility away from a duty toward other people, and toward responsibility to the law. The ideal of justice derived through obedience to moral laws underpins Walzer's theory of war. As discussed in the first chapter, Walzer presents his articulation of the moral rules of war under the heading of the War Convention, 'a set of articulated norms, customs, professional codes, legal precepts, religious and philosophical principles, and reciprocal arrangements that shape our judgements of military conduct' (2006a: 44). Crucially, Walzer argues that acting in accordance with the War Convention satisfies ethical responsibility in relation to war: 'So long as they [combatants] fight in accordance with the rules, no condemnation is possible' (2006a: 128). In other words, adherence to the War Convention resolves all questions of ethical responsibility and justice in war. Walzer describes the War Convention as a set of maximal rules that faithfully represent the minimal laws of war. This minimal dimension endows the War Convention with universal jurisdiction; it is therefore the most direct embodiment of Walzer's thick and thin morality – a particularist discourse underpinned by universal values.

Walzer frames the universal aspect of the War Convention in terms of the language of rights. Importantly, Walzer founds the possibility of justice in war upon a strict defence of the rights of life and liberty: 'For the theory of justice in war can indeed be generated from the two most basic and widely recognised rights of human beings – and in their simplest (negative) form not to be robbed of life and liberty' (1983: xv). Walzer assures us that the rights of life and liberty should be understood, in the context of war, as 'something like absolute values' (2006a: xxiv). While the term 'absolute' suggests that these rights are fundamentally inviolable (people cannot be robbed of life and liberty), the prefix 'something like' indicates that rights may be conditional in particular circumstances. The most direct condition placed on the absolute right to life is intimately related to the killing of combatants in war. Walzer acknowledges that the death of combatants is an unavoidable reality of war. Nevertheless, he argues that it is in distinguishing the killing of combatants from the killing of non-combatants, that we found the basis for the War Convention:

> 'Soldiers are made to be killed,' as Napoleon once said; that is why war is hell. But even if we take our standpoint in hell, we can still say that no one else is made to be killed. This distinction is the basis of all the rules of war.
>
> (Walzer 2006a: 136)

Although Walzer accepts that combatants are 'made to be killed', he simultaneously maintains that this loss of life is not tantamount to the violation of absolute rights. Instead, Walzer claims that combatants forfeit their rights in the context of war (2006a: 137). In this respect, Walzer wants to stress a conception of justified killing in terms of forfeiture rather than in terms of ethical sacrifice. Absolute rights can be forfeited without sacrificing ethical responsibility; and he justifies combatants' forfeiture of rights by arguing that rights are lost through individual actions:

> A legitimate act of war is one that does not violate the rights of people against whom it is directed. It is once again, life and liberty that are at issue . . . I can sum up their substance in terms I have used before: no one can be forced to fight or to risk his life, no one can be threatened with war or warred against, *unless through some act of his own* he has surrendered or lost his rights.
>
> (Walzer 2006a: 135, italics mine)[1]

Combatants thus cannot forfeit their rights unless the forfeiture is the result of their own freely taken actions. On the opposite side of the dichotomy, Walzer describes non-combatants as innocent and therefore immune from intentional attack.[2] Walzer elaborates that innocent people have done nothing, and are doing nothing, that would entail the forfeiture of their rights (2006a: 146). Again, the central point posited is that immunity from attack in war is directly connected with individual actions, and individuals disengaged from military activities cannot be attacked (2006a: 43). The link between rights forfeiture and individual action is clear: it is only when an individual actively engages in the military effort that they can be legitimately killed during wartime. Walzer's understanding of non-combatant immunity is, in turn, a key component of contemporary Western rules of war fighting. For example, the US Army and Marine Corps Counterinsurgency Field Manual identifies protecting non-combatants as the primary goal of counterinsurgency (COIN): 'securing the civilian, rather than destroying the enemy, [is] their [COIN troops] top priority' (Petraeus 2007: xxv). Thus, the theoretical distinction between combatants and non-combatants is often viewed as the defining aspect of ethical responsibility in war: to be ethically responsible is tantamount to refraining from targeting non-combatants. However, this distinction also implies that the killing of combatants is morally justified and detached from ethical considerations.

The purpose of this chapter is to illustrate, within the context of the Iraq War, why the conventional understanding of non-combatant immunity is both logically inconsistent and practically impossible. This chapter, in other words, seeks to explain the ethical ambiguities retained in Walzer's foundation of the moral rules of war. Highlighting the relationship between ethics as response and war fighting in Iraq, this chapter argues that the killing of combatants in Iraq constitutes an ethical sacrifice. In Walzer's terms, rights are not merely forfeited; combatants' rights are sacrificed in the name of ethical responsibility toward non-combatants. In destabilising the presupposed basis of the War Convention this chapter points toward the inadequacy of Walzer's conception of wartime morality. In response, I posit an understanding of ethics in which responsibilities are produced and reproduced within the specific contexts of individual wars.

Identifying the target

This chapter demonstrates Walzer's inability to maintain the combatant/non-combatant distinction. However, the War Convention also fails to protect non-combatants even when the distinction is unproblematically accepted. The most immediate difficulty in protecting 'the innocent' is the fact that combatants are not always identifiable as such. For example, in Iraq some Ba'athist fighters during the initial invasion in 2003, and practically all resistance fighters since the fall of the Ba'ath regime, have refused to visibly identify themselves, to US troops, as combatants. There are, of course, clear strategic reasons for this refusal; primarily, if resistance fighters fought a conventional war against the largest and best equipped military in the world, they would undoubtedly be defeated. What is particularly important to the present discussion, nonetheless, is how the US Military responded to this problem: how they attempted to separate legitimate objects of attack (combatants) from the general population (non-combatants). Walzer addresses this specific problem in his discussion on guerrilla war.[3] While he recognises the moral impetus of a defeated population to endeavour to regain control of their homeland, he argues that resistance to occupation is punishable by death: 'If citizens of a defeated country attack the occupation authorities . . . [it is] a breaking of political faith, punishable, like ordinary treason of rebels and spies, by death' (2006a: 177). Again, this is a restatement of individual forfeiture: if resistance fighters choose to attack the occupying forces, they forfeit their right to life. Yet the most pressing problem, for Walzer, is that by refusing to openly identify themselves as combatants, resistance fighters threaten the foundation of the War Convention, making it impossible for enemies to accord combatants and non-combatants their distinct privileges (2006a: 180). In fact, Walzer concedes that resistance fighters' unwillingness to identify themselves as combatants often means that additional risks are necessarily imposed on ordinary people (2006a: 178). In short, resistance fighters attack from within the population and therefore bring the risk of violent response upon civilians.

The US occupation of Iraq provides clear examples of the problems faced by troops attempting to differentiate between resistance fighters and civilians, and the COIN manual often discusses the difficulty in separating resistance fighters from the general population (Petraeus 2007: xxv, 17, 92). The manual underscores the main problem by stating that occupied populations often drift between the roles of resistance fighter and civilian follower (Petraeus 2007: 22). Therefore, it is hard to identify resistance fighters because resistance fighters are practically indistinguishable from non-violent civilian populations hostile to occupying forces. This point is emphasised by journalist Dexter Filkins, who maintains that although resistance fighters sometimes fought, for the majority of the time they were just standing around like every other Iraqi (2009: 122). This, of course, presented a major problem to US troops charged with protecting Iraqi civilians and catching or killing resistance fighters. In the early stages of the occupation, the main way that US troops attempted to unearth resistance fighters was the mass detention of Sunni males who were assumed to represent the core of Iraqi resistance. The primary problem encountered by US troops in Iraq was that civilians were largely unwilling to provide information about the resistance; therefore, the US knew very little about resistance activities, composition and structures. To rectify this information gap, the US imprisoned thousands of Iraqis in an attempt to generate 'actionable intelligence' on resistance activities. Independent reporter Dahr Jamail describes how US troops responded to a resistance attack by cordoning off several blocks of a Sunni neighbourhood, conducted house-to-house searches and 'carted off' all the males to prison (2008: 75). The number of people detained in line with this policy was staggering: 30,000 to 40,000 Iraqis were officially held in US detention facilities during the first eighteen months of the occupation (Ricks 2007: 199). The US efforts to gather intelligence on the resistance, however, proved largely unsuccessful, despite the mass detentions. In this regard, it would appear that Sunni populations were, by and large, unwilling to inform on resistance fighters.

For Walzer, public support is the key deciding factor in the status of resistance movements. According to Walzer, if the resistance is not supported by the people, the people will give fighters up to the occupying authorities. However, if the people generally support the resistance they will not give them up and there is no way of separating them from the population (2006a: 186–196). If a military can no longer separate combatants from the general population, they can no longer fight without directly targeting civilians and the war can no longer be justified. In this sense, public support for resistance movements is intimately tied to Walzer's understanding of legitimacy. His central argument is that mass popular support transforms resistance into a force of legitimate self-determination; it makes the resistance the legitimate rulers of the country and the war against the resistance can no longer be fought justly (Walzer 2006a: 196). Hence, Walzer wants to suggest that when a resistance movement reaches a certain level of popular support it is representative of communal popular will. Again, this is tied to his ideal of community, and his belief that a population's willingness to engage in violence symbolises a commitment to fight for self-determination. Walzer conceives guerrilla warfare in a two-dimensional way: it is a battle between the internal government/occupying authority and the resistance fighters, and public support is the prize. But resistance in Iraq cannot be characterised in this way. First, there were multiple resistance movements with divergent, often conflicting, goals – Sunni resistance, for instance, was often pursuing goals counter to those of Shi'a and Kurdish communities. Second, and more importantly, resistance support was in part fostered through fear of violence; in other words, if civilians tried to give up the resistance, like Walzer suggests, they risked being targeted by local fighters. In this respect, the communities supposedly harbouring resistance fighters were assailed from both sides, fearing US detention and resistance retribution. Filkins argues that civilians' main objective was to stay in the good graces of both sides because they had to live in the neighbourhoods after US troops had gone (2009: 115–122). Therefore, it is in the context of fear – rather than a popular drive to self-determination – that the inability to separate resistance fighters from the general Iraqi population should be understood. In this sense, by occupying Iraq, US troops immediately put the lives of ordinary civilians at risk. Civilians who had, in Walzer's terms, done nothing that would entail the forfeiture of their rights were faced with potential dangers from all corners.

Combatant rights

The difficulty in identifying combatants in the context of Iraq highlights a significant problem in adapting Walzer's theory: the problem of universalising rules and norms abstracted from the specific contexts to which they are intended to be applied. More specifically, the strategic refusal of resistance fighters to openly identify themselves as combatants in Iraq fundamentally blurred the lines between combatants and non-combatants. Nonetheless, Walzer's principle of non-combatant immunity is primarily geared toward conventional war fighting, and so appears far more applicable to US troops who generally identified themselves as active combatants in Iraq. Walzer's justification of the killing of combatants explicitly links the act of soldiering to an individual's loss of rights: 'The immediate problem is that the soldiers who do the fighting, though they can rarely said to have chosen to fight,[4] lose the rights they are supposedly defending' (2006a: 136). The soldiers fighting to defend a community's rights thus forfeit their own individual rights in the name of the collective defence. Walzer claims that by forfeiting their civilian rights, combatants gain a new set of rights and obligations – primarily the right to kill enemy combatants (2006a: 40–41). He argues that soldiers gain war rights that are grounded on what he calls the moral equality of combatants. For Walzer, moral equality means that combatants can kill without the act of killing constituting murder: '[n]either man is a criminal, and so both can act in self-defence. We call them murderers only when they take aim at non-combatants, innocent bystanders (civilians), wounded or disarmed soldiers' (2006a: 128). In Walzer's argument, the killing of combatants does not entail a violation of the absolute right to life and, therefore, does not compromise the integrity of the moral rules of war.

The idea of combatant rights is essential for Walzer's separation of *jus in bello* (justice in war) from *jus ad bellum* (just recourse to war).[5] Walzer needs to maintain the categorical separation of these two aspects of warfare, i.e. one can fight an unjust war justly and vice versa. Indeed, the division between different spheres of justification was a key aspect of Walzer's defence of US post-war reconstruction in Iraq. Walzer accomplishes the *ad bellum/in bello* separation by differentiating between the moral responsibility of political leaders to ensure that recourse to war is taken in line with just war criteria, and the responsibility of the combatants to fight in accordance with the War Convention.[6] Walzer's argument is that the crime of war is the specific crime of the political leader(s) of an aggressor state. In turn, because combatants do not decide to start the war they are absolved of the crime of aggression. As such, both aggressor combatants and resisting combatants face each other as mutually innocent of the crime of war and are, therefore, morally equal.[7] Walzer's description of moral equality provides an interesting contrast to his understanding of the forfeiture of rights. People forfeit rights on the basis of identifiable individual actions; yet moral equality is granted to all combatants as a homogenous collective regardless of individual actions or culpability. In this respect, Walzer's conception of combatant rights provides an example of what Derrida describes as the individualisation of the role rather than the person; 'the objective or quantifiable equality of roles not persons' (2008: 37). Walzer thus treats moral equality as an undifferentiated reality applicable to any individual categorised as a combatant. Crucially, he never asks if individual combatants actually accept the right to kill or be killed in the terms that the War Convention suggests.

Walzer's assumed undifferentiated reality is challenged by the different expectations of combatants employed in different roles within the US Military. For example, those trained for roles in medical or reserve units of US forces often view killing as something alien, and in some cases counter, to the ethos of their vocation (Gutmann and Lutz 2010). In fact, many disenchanted US troops argue that they only joined the services in light of promises made by recruitment officers that they would never have to serve in a war zone or be expected to kill (Mejia 2008: 15). On the other hand, US infantry troops serving in Iraq generally appeared to be more comfortable with their duty to kill. Marine officer Nathanial Fick provides a reflective analysis of killing the enemy in terms of moral equality:

> I found no joy in looking at the men we'd killed, no satisfaction, no sense of victory, or accomplishment. But I wasn't disturbed either. I fell back to an almost clinical detachment. The men were adults who chose to be here. I was an adult who chose to be here . . . the fight was fair.
>
> (Fick 2007: 273)

Although Fick's reflections mirror the logic of the War Convention,[8] the more blunt assertion of British sergeant Dan Mills perhaps gets closer to the general attitude of front line infantry: 'He was the enemy, and all I gave a shit about was that he was dead' (Mills 2008: 73). Mills is quick to point out that his satisfaction at the death is derived from a sense of professional duty to eliminate threats to his comrades, and this ideal of professionalism is largely in line with US accounts. For example, Evan Wright was told by the Marine chaplain that Marines often come to him for counselling if they haven't fired their weapons, because they feel guilty for not doing their job properly and failing to protect their comrades (2005: 239). What is interesting about such attitudes to killing, in the context of this discussion, is that while the killing of non-combatants is deemed morally problematic in US Military doctrine, the parallel killing of enemy combatants is rendered as non-moral. Indeed, the targeting of identifiable combatants is not considered to be a question of morality at all, but one of professionalism. In short, good soldiers engage the enemy in order to protect their fellow troops. In this sense, combatants are acting out of professional duty rather than moral responsibility: the ethical relationship implied in the concept of moral equality is submerged within another discourse on what constitutes professional military conduct, and adherence to the laws of military conduct replaces responsibility to other people.

US infantry troops in Iraq appeared to willingly accept their professional duty to kill the enemy. However, the ancillary requirement, stipulated in the War Convention, that combatants accept the equal right of the enemy to kill them proves far more complex. On the one hand, US troops are meticulously prepared for the possibility that they, or their comrades, may die in battle. In fact, the acceptance of the possibility of death for US combatants is regimented in both bureaucratic[9] and symbolic[10] ways. On the other hand, US combatants' acceptance of the risk of death is at best met with conditional approval. The most clearly elucidated condition is that combatants' lives are not sacrificed cheaply. Fick explains: 'My Marines and I were willing to give our lives, but we preferred not to do so cheaply. The fear was a realisation that my exchange rate wasn't the only one being consulted' (2007: 236). Implicit within Fick's account is the idea that US troops are willing to sacrifice their lives, but only for the right reasons. In the context of Iraq, the primary reason presented for the sacrifice of US lives was the defence of freedom: freeing Iraqis from Ba'athist rule and thereby ensuring that the US would not be attacked by terrorists or Saddam Hussein's WMDs. The sacrifice of US combatant lives was justified because it would ensure that Iraqis were free from tyranny and that the US would be free from external threats. Fick's fear, however, that other factors were involved was one echoed by a number of US combatants. Troops were afraid that they were being placed in danger because of their government's imperial ambitions (Gutmann and Lutz 2010: 49), because superiors wanted to win medals and promotions (Mejia 2008: 174) and even because of command incompetence (Wright 2005: 428–433). In short, US combatant acceptance of sacrifice in Iraq was dependent upon context: why their lives were being risked for a specific purpose in a specific instance. This brings us toward a more nuanced and conditional account of combatant perceptions of moral equality. Moral equality is not simply accepted as a de facto principle; rather, the context in which combatants' lives are risked is important in regard to combatants' acceptance of the imperative. The Derridean concept of *se donner la mort* helps us understand combatants' acceptance of moral equality: acceptance of the risk of death is dependent upon what the sacrifice means – and to what end it is oriented.

While Walzer would undoubtedly support a more reflective military attitude that takes context into account, this nevertheless presents a distinctive problem. Given the necessity for political leaders to impress the justness of their cause, it is hardly surprising that combatants are often convinced that their enemies are fighting for amoral causes.[11] In Iraq, the Ba'athist regime and resistance fighters were often explicitly referred to as intrinsically amoral. In 2004 US Deputy Secretary of Defence Paul Wolfowitz, for example, claimed that the Ba'athists were as bad, or worse, than the Gestapo (cited in Ricks 2007: 386). Derrida argues that war is always a battle over the meaning of justice, with warring sides fighting for the right to declare their causes wholly moral (2008: 86–87). Derrida explains that each side strives to present their violence and sacrifices as morally justified, and one of the easiest ways to achieve this is by depicting the enemy as vile murderers. In the context of Iraq, successive US governments have gone to great lengths to emphasise the barbaric nature of the Ba'ath regime and the despotic character of Saddam Hussein. The net result of the moralisation of the war was that US forces entered Iraq viewing their Iraqi adversaries as wholly unjustified in their resistance of US attempts to 'liberate' Iraq from tyranny.[12] Under such conditions, it was difficult for combatants to share a sense of moral equality with those who, they were persistently told, were working toward immoral ends.[13]

Perceptions of Ba'athist immorality, combined with Iraqi responses to occupation, fuelled confusion and anger among US troops. Troops were sold the idea of intervention under the premise that they were rescuing Iraqis from a despotic regime, and were assured that Iraqis would welcome them as liberators. Yet, when the dust of the demise of the Ba'ath regime settled, troops faced local distrust and violent resistance. US troops' sense of dejection and betrayal was reinforced by Iraqi responses to violence, with US troops often witnessing scenes of public jubilation when their comrades were killed (Ricks 2007: 329). Captain Oscar Estrada poignantly sums up the frustrations felt by US troops who found themselves in the midst of a violent resistance:

> I think of . . . the children who burst into tears when we point our weapons into their cars (just in case), and the countless numbers of people whose vehicles we sideswipe as we try to use speed to survive the IEDs [improvised explosive devices] that await us each morning. I think of my fellow soldiers and the reality of being attacked and feeling threatened, and it all makes sense – the need to smash their cars and shoot their cows and point our weapons at them and detain them without concern for notifying their families. But how would I feel in their shoes? Would I be able to offer my own heart and mind?
>
> (Cited in Ricks 2007: 365)

As Estrada highlights, US troops found themselves torn between the purported humanitarian intentions of their mission and the daily threat they faced from the people they were supposed to be helping. The targeting of US troops is interesting in that it simultaneously shook and reinforced combatants' willingness to sacrifice. Troops were shaken by the realisation that their comrades were dying in the name of a people who did not want a US presence in their country. However, the targeting of combatants reinforced troops' commitment to honour the sacrifice of their fallen comrades. Reporter David Finkel describes the confusion within troops' minds: 'Three dead . . . this is exactly why we need to get out of Iraq, to honour the sacrifice, and this is why we need to stay in Iraq, to honour the sacrifice' (2011: 129). In this sense, combatant sacrifice, to a certain extent, became a means of its own justification: further sacrifice became necessary to render the loss of comrades' lives meaningful. Good men and women had died and the job had to be finished so that they had not died in vain. Nonetheless, this is not conducive to moral equality between combatants. Instead, it provides an indication that US troops believed that their lives were worth more than their adversaries' lives.[14] For example, Finkel recounts graffiti scrawled upon a locker at a US military base in Iraq: 'No Iraqi man, woman, or child is worth one drop of an American soldier's blood' (2011: 161).

US combatants in Iraq generally accepted that their lives could be risked by their nation and, in some cases, the acceptance of risk was dependent on the perceived moral character of the war being fought. However, US combatants did not necessarily accept that it was justified for them or their comrades to be killed at the hands of their purportedly immoral enemies. Moral equality, therefore, is not tantamount to a practical reality in the way Walzer describes. His ideal of moral equality can only work in the context of a predefined role abstracted from subjectivity. It does not take account of the ways in which individual combatants interpret this role. In the context of Iraq we are presented with a far more fluid relationship between US troops and moral equality. While Walzer argues that combatants' right to life is de facto forfeit, individual combatants do not necessarily subscribe to this blanket revocation, especially in terms of US combatants' acceptance of resistance fighters' right to kill them. Walzer contends that combatants' unconditional acceptance of the War Convention ensures the fulfilment of justice. Yet the practicalities of war fighting in Iraq illustrate how the context of the war directly impacts upon the ways in which general principles are interpreted. Non-combatant immunity proved impossible to rigidly enforce in Iraq due to the nature of guerrilla warfare; and US combatants view the principle primarily as a professional rather than inherently moral duty. In turn, moral equality is problematic because US combatants fighting in Iraq did not necessarily accept resistance fighters' equal right to kill. Ultimately, we are faced with contextual negotiations of Walzer's principles, not strict adherence.

Justifying the loss of rights

The last section looked at the difficulties in practically implementing the War Convention in Iraq. However, Walzer's theoretical justifications for the forfeiture of rights are problematic in and of themselves. He offers two primary justifications for combatants' loss of rights:

> Simply by fighting, whatever their private hopes, and intentions, they have lost their title to life and liberty, and they have lost it even though, unlike aggressor states, they have committed no crime.
>
> (Walzer 2006a: 136)

> He can be personally attacked only because he is already a fighter. He has been made into a dangerous man, and though his options may have been few, it is nevertheless accurate to say that he has allowed himself to be made into a dangerous man. For this reason he finds himself endangered.
>
> (Walzer 2006a: 145)

Walzer explicitly states that soldiers are destined for dangerous places (2005: 73) and argues that the 'slaughter' of combatants should be viewed as moral (2006a: 40); in other words, war necessitates risks to combatants' lives and these risks are deemed to be morally acceptable. Remembering that Walzer's central argument is that rights are forfeited on the basis of individual actions, it is important to analyse these two justifications, paying specific attention to the acts committed by combatants that have resulted in their loss of rights.

Simply by fighting

Walzer's first justification revolves around the act of fighting. Because combatants are in the business of fighting, they can legitimately be attacked in self-defence. However, this justification is almost immediately placed in a tenuous position when Walzer asserts that soldiers do not regain their rights simply by not fighting, i.e. if an individual combatant chooses not to engage the enemy they remain a justifiable target (2006a: 138). Again, this implies that the designated role of the combatant – and not the individual choice to actually fight – is the determining factor in the relinquishment of the right to life. Walzer argues that there are two instances in which combatants regain their right to life: if they are captured by the enemy (although their right to liberty remains revoked), or if they are wounded on the battlefield. Walzer maintains that combatants are liable to attack if they return to the warzone. Immunity – and the associated return of the right to life – in this sense, is only a temporary respite between sacrificial deployments.[15] Therefore, in Walzer's formulation, combatants are liable for attack whenever they are present on the battlefield and capable of fighting. As such, Walzer's first justification needs to be re-stated as: 'Simply by fighting in *the first instance* . . . they have lost their title to life and liberty *so long as they remain on the battlefield.*' This brings us to the question of how soldiers come to fight in the first instance, a question that will be addressed later in this chapter.

Thus far we have accepted Walzer's conception of the right to life as a simple negative barrier: a person cannot be killed unless, through some act of their own choosing, they have forfeited their right to life. Yet an important ethical problem emerges in Walzer's account if we consider the right to life as something more than merely the negative prohibition on killing the innocent. Walzer attempts to fix the loss of rights within the confines of the battlefield; soldiers are at risk only so long as they remain fighting. Nevertheless, in order to maintain this distinction, Walzer must discount the long-term ramifications of war fighting that soldiers bring home from the battlefield. In contrast, ethics as response suggests that we must account for the wider consequences that derive from the decision to send men and women to kill and be killed in war. The most obvious examples of the way in which war changes combatants' lives are cases when troops are physically injured. The official estimate of US troops injured in Iraq, as of August 2013, totals 31,943.[16] As striking as this number is, what is of more concern, for this discussion, are the practical implications of life for injured soldiers. Finkel paints a horrific picture of the extent of injuries suffered by US troops returning from Iraq:

> He put on a protective gown, protective boots, protective gloves and walked toward a nineteen-year-old soldier whose left leg was gone, right leg was gone, right arm was gone, left lower arm was gone, ears were gone, nose was gone, and eyelids were gone, and who was burned over what little remained of him.
>
> (Finkel 2011: 201)

While Finkel's example undoubtedly depicts an extreme case of combatant injuries, it nonetheless serves as an important reminder that the lives that soldiers leave before deploying are not necessarily the same lives that they return to. In other words, Walzer's negative barrier does not take account of the life that combatants are capable of living after their experiences of war, or how injuries to combatants affect family members and other members of society.

The transformation of US combatants' lives is not solely restricted to physical injuries. A significant number of US troops returning from Iraq suffers from mental and emotional traumas. The extent of the psychological impact upon US troops was underscored by the revelation in 2011 that more active-duty US soldiers committed suicide that year than had died in combat (Pilkington 2013). In this sense, the psychological costs of war expand far beyond the battlefield. The clinical aspect of psychological injuries is often studied under the heading of post-traumatic stress disorder (PTSD). Symptoms of PTSD are commonly believed to include depression, anger, uncontrollable anxiety, survivor's guilt, re-living trauma via vivid memories, risk avoidance and a number of other psychological problems (Litz *et al.* 2009, Hodge *et al.* 2004). In short, PTSD severely limits sufferers' ability to live their day-to-day lives. In terms of Iraq veterans, a 2008 RAND study found that at least 19 per cent of US troops – over 300,000 people at that stage of the war – returned home suffering from symptoms of PTSD (Gutmann and Lutz 2010: 141).[17] Iraq has proved a fertile ground for problems like PTSD, partly because US troops felt they were constantly at risk of attack. A soldier explains the unique conditions of Iraq to Finkel: 'In other wars, the front line was exactly that, a line to advance toward and cross, but in this war, where the enemy was everywhere it [the front] was anywhere out of the wire' (2011: 35). The account provided by this soldier aligns with Patočka's phenomenology of warfare (1996: 119–137). In his analysis, Patočka describes the front as an overwhelming vertigo of human experience resulting from the proximity of the self to its own death. Patočka also stresses the necessity for soldiers to break from the front in order to recover from the emotional intensity of living on the border of death. For US troops in Iraq, the experience of the front became a daily routine of street patrols that confronted soldiers with the constant threat of IEDs, rocket attacks and gun fire. Even in the relative safety of compounds troops still faced the risk of unpredictable mortar attacks. Lengthy deployments and the ever present possibility of redeployment, exacerbated by the practice of Stop-Loss, drastically increased the psychological toll on US combatants. Finkel's narrative offers some very informative discussions on the implications of the exposure to the pervasive Iraqi front. One soldier talks directly about the daily prospect of patrols: 'Every time I go out on patrol, I feel sick, it's like, I'm going to get hit, I'm going to get hit, I'm going to get hit' (cited in Finkel 2011: 116); while another soldier descended into nihilistic depression, seeking dangerous jobs and extra risks in the hope he would be killed: 'Bottom line – I wanted it over as soon as possible, whether they did it or I did it' (cited in Finkel 2011: 188).

Psychological and physical traumas are not simply abandoned to the battle-field like the vitality of the dead. They are dragged back into the civilian lives of US troops, and the lives of their families and friends. The impact of these traumas has been quite dramatic, with Iraq veterans experiencing higher rates of divorce, unemployment, homelessness, suicide and domestic abuse (Gutmann and Lutz 2010: 190). War trauma has clear consequences for combatants' families, friends and extended communities. The first impact is that thousands of US families have lost a loved one during the Iraq War.[18] The loss of a loved one is poignantly outlined by the wife of a twenty-two-year-old serviceman killed in Iraq: 'He was my everything, and he was ever since the day I met him. My heart, my soul, my friend, my husband' (cited in Ricks 2007: 364). Others, however, suffer the direct effects of service people returning from war with physical and psychological traumas. While the impacts of physical injuries are more visible, with physical limitations or the necessity for full-time care, psychological injuries also have major implications for the families of veterans. In Demers's (2009) study of the impact on veterans' families, the most persistent response she encountered was that the people who returned home from Iraq were fundamentally different from those who had departed. As one female participant stated: 'His anger is toxic, and I'm tired of fighting. I'm tired of watching him drink himself to sleep night after night, and I'm tired of being his punching bag . . . He's turned into a drunken monster, and I don't know what to do' (cited in Demers 2009: 4); while another simply asserted, 'nobody comes back the same' (ibid.). The belief that war has transformed or deformed identity is particularly widespread among Iraq veterans. US combatants fighting in Iraq were often plagued by the fear that the war had transformed them into someone else, or stripped them of their humanity. For example, a Marine in Wright's narrative felt he had lost his last shred of humanity (2005: 281); and former army reservist Carlos Mejia referred to his experience of Iraq as the many deaths of the soul (2008: 213). In this regard, Finkel's account of a seminar given to troops returning home is, perhaps, most telling:

At the chapel, there was a mandatory seminar on what to expect in the months ahead. It's normal to have flashbacks, the soldiers were told, normal to have trouble sleeping, normal to be angry, normal to be jumpy – and didn't that make everyone feel better.

(Finkel 2011: 238)

US troops are, therefore, officially told that they should expect to feel different, they should expect to be disturbed, they should expect to feel abnormal – and feeling like that is perfectly normal. In short, there is a tacit acknowledgment that war is an adversely transformative experience with clear consequences for combatants and their families.

The experiences of US combatants returning from Iraq illustrates why ethical responsibility toward combatants is not adequately addressed via the understanding of the right to life and liberty stated in terms of negative prohibitions. Walzer's argument suggests that combatants must risk their lives for the duration of the conflict so that the collective rights of the community can be protected. However, military service cannot be understood as a temporary sacrificial vocation. Instead, war produces new modes of existence for those who experience its traumas. Returning combatants may retain a right not to be killed, but they have, in certain ways and certain respects, lost the life they had prior to service. Importantly, this loss of prior life is intimately related to Walzer's understanding of communal life and maximal morality. He presents human existence in terms of the shared life built together by members of a community. It is precisely this conception of life that Walzer argues necessitates violent defence in times of war. Yet simultaneously, combatants who survive the trauma of war risk losing the shared existence they have built prior to deployment. Stated in another way, Walzer justifies war in defence of self-determination, but those enlisted to fight the war risk losing their sense of self. Our ethical responsibilities to combatants, therefore, cannot be conceptualised in terms of the temporary forfeiture of the right to life, because war itself is constitutive of the referent. War transforms the lives of those who are exposed to it and as such, ethical responsibility needs to be understood in the frame of de-limitation: long-term consequences rather than short-term risks. The risks imposed on combatants are projected into their future lives; the experience of war affects the ways in which returning combatants relate to themselves and others. The consequences of asking men and women to fight, as illustrated in the preceding examples, are not limited to the battlefield, because they have ethical impacts far beyond this scope. Hence the first justification for the forfeiture of combatants' rights is subject to a perpetual deferral: we cannot justify the forfeiture without taking account of the longer-term direct and indirect consequences of sending men and women to kill and be killed on the battlefield.

Danger and threat

Walzer's second justification rests on the assertion that the combatant is a 'dangerous man' and, further still, that he has allowed himself to be made into a dangerous man.[19] Bracketing off the question of the *allowed* for a moment, a question that is implicitly linked to the soldier fighting in the first instance, let us turn our attention to the question of what it means to be dangerous. The principle of non-combatant immunity depicts *all* combatants as dangerous. Nevertheless, there are questions about the extent to which military medics, cooks and supply convoy drivers can be considered dangerous in Walzer's terms. While these roles are integral to the military logistics machine, Walzer argues that civilians who contribute to the military effort can only be attacked when they are directly contributing. When discussing workers of munitions factories and others who help produce what militaries need to fight, Walzer is quite clear that such people can only be attacked when they are engaged in threatening activities: 'These are not armed men, ready to fight, and so they can only be attacked in their factory . . . when they are actually engaged in activities threatening and harmful to their enemies' (2006a: 146). In a similar manner, non-infantry personnel are not armed and trained to fight in the way Walzer suggests all combatants are. Yet it is impossible to limit the risk posed to non-infantry personnel to working hours when they live among those who are 'ready to fight'. As such, it is evident again that the War Convention rests upon an assumption about the threatening nature of the category of combatant, rather than an assessment of individual actions. Another iteration of this problem that has emerged in Iraq can be found in the risks posed to non-military civilians working alongside US troops. For example, local interpreters were crucial to the US efforts to rebuild Iraq as they were the primary means through which the military could engage with the local population. Finkel argues that Iraqi interpreters were required to take on soldiers' risks, while also risking being branded as pariahs by their neighbours, for a salary well below that of US troops (2011: 159). The threat of being viewed as a pariah is particularly significant to Walzer's conception of danger as it illustrates how the risks inflicted on interpreters could not be simply limited to working hours. As journalist Patrick Cockburn claims, Iraqis living in a destitute economy often had to choose between permanent unemployment, immigration, or risking their lives by working with US troops (2007: xx).

The predicaments faced by non-infantry troops and civilians working and living alongside infantry combatants highlights the simplistic nature of Walzer's uniform classification of combatants as dangerous men. If Walzer wants to justify the revocation of combatants' right to life and liberty on the grounds of the identifiable actions of individuals, he needs to provide some mechanism for separating threatening combatants from their non-threatening, or at least not immediately threatening, comrades and colleagues. Otherwise, Walzer risks providing a blanket justification for killing without any judgement of individual culpability. Walzer attempts to qualify combatant forfeiture by asserting that the threat posed by combatants conditions our response (1994: 24); he argues that it is only when the soldier tries to kill me that 'he' alienates 'himself' from me and our common humanity, thereby forfeiting 'his' right to life (2006a: 144). Following this, we would assume that it is the direct threat posed by the individual combatant in a specific instance that triggers the forfeiture of rights; combatants can be targeted in particular contexts where the threat they pose is evident. Yet Walzer rejects this model by stating that 'the threatening character of soldier's activities is a matter of fact' (2006a: 200). Once again, Walzer emphasises the argument that combatants can be attacked because of the assumed threat attached to their role, rather than their individual actions. Interestingly, while Walzer designates combatants as a de facto threat, he goes to great lengths to avoid similar blanket statements in regard to other aspects of his theory of war. For example, when discussing the concept of pre-emptive war, Walzer makes a direct attempt to define the parameters of justified threat. Understanding that claims of threat can be employed fallaciously for strategic ends, Walzer signals his intention to define non-arbitrary standards of what it means to be threatened; to this effect, he claims that there is an objective standard of 'just fear':

> I can only be threatened by someone who is threatening me, where 'Threaten' means what the dictionary says it means: 'to hold out or offer (some injury) by way of threat, to declare one's intention of inflicting injury'.
> (Walzer 2006a: 78)

Walzer expands upon this definition by arguing that a violent response is justified only when the threat is evident in some material sense (2006a: 80). In other words, we must wait for some wilful act of the adversary before we can appeal to the objective standard of just fear. Walzer concludes this discussion by stressing that 'the idea of being under threat focuses on what we had best call simply *the present*' (2006a: 81, original italics). He claims that the objective standard of just fear helps us to distinguish between those who can be described as present instruments of an aggressive intention and those who may represent a distant danger (2006a: 80). In short, an objective threat is when a material offering of injury is declared and intended in the present.

How then does this conception of threat relate to the assumed danger posed by combatants? Following from Walzer's definition, we would assume that a soldier can only be attacked if they are presently offering a clear intention to kill or injure their adversary. However, this is a principle that Walzer adamantly rejects when he presents us with the figure of the naked soldier, as recounted by Robert Graves:

> I saw a German, about seven hundred yards away, through my telescope sights. He was taking a bath in the German third line. I disliked the idea of shooting a naked man, so I handed the rifle to the sergeant with me. 'Here, take this. You're a better shot than I am.' He got him; but I had not stayed to watch.
> (Cited in Walzer 2006a: 140)

Walzer stresses the moral dilemmas involved in killing soldiers who are not presently engaged in acts of warfare. However, he concludes by definitively asserting that the killing of the naked soldier is justified (2006a: 143). As such, the definition of threat applicable to combatants is detached from their present action; it is based upon their past actions, that they became soldiers, and their assumed future actions, that they will injure (or at least aim to injure) their adversaries in the future. The assumption that combatants will act dangerously in the future belies Walzer's depiction of a material offering of threat. Further to this, an assumed future threat could be applied to any number of non-combatants; for example, in the context of the volatile occupation of Iraq, all civilians could potentially join or assist resistance movements, meaning that any civilian could be depicted as a potential future threat. Nonetheless, Walzer fervently rejects targeting civilians on the basis of potential threat for this reason (2006a: 214). Again, we see that the presupposition that there is something inherently dangerous about combatants is central to Walzer's justification for the forfeiture of rights: the role of combatant designates a dangerous person and this justifies the unproblematic revocation of their absolute right to life.

Although I have explained why Walzer's second justification rests upon the assumed threat posed by combatants, this still implies that if threat was aligned with the present actions of an individual it would resolve the moral problematic of killing in war. What I would like to do now is to illustrate why Walzer's conception of objective threat determination is unconvincing. Walzer requires determinations of threat to be made in the present; yet this ideal of objective threat determination is called into question by the prevailing uncertainty that infuses the actual contexts in which determinations are made. Looking at the example of roadblocks in Iraq will help illustrate this tension between determinations of threat and uncertainty.

Roadblocks represented a day-to-day security activity of coalition troops in which temporary barriers, often constructed via concertina wire, were erected in an effort to temper the movement of resistance fighters throughout Iraq and prevent car bombings. During the initial stages of the war, roadblocks functioned under the auspices of preventing Fedayeen and other Iraq army troops from fleeing defeated towns and, thereby, re-commencing the armed resistance in another location (Wright 2005: 278–79). However, during the occupation the onus swiftly shifted to stopping the movement of resistance fighters and weaponry, and halting the increasingly lethal tactic of car bombing. (Ricks 2007: 215–216).[20] Roadblocks in Iraq can generally be described as points at which US military vehicles set up barriers that civilian vehicles were not allowed to breach.[21] Roadblocks also provide a pertinent example, in another sense, because they allowed for conventional planning and therefore should have provided troops with clear templates for determining threats and acting appropriately. The US Marine guidelines for roadblocks state that Marines should fire warning shots at the approaching vehicle.[22] If the vehicle continues to approach, the soldiers are required to shoot directly at it until it stops (Wright 2005: 278–79). The COIN manual highlights roadblocks as a key instance in which combatants' responsibility to protect civilians is placed in direct conflict with their responsibility to protect their fellows: 'Checkpoints are perhaps the most vivid example of the cruel trade-offs pushed down to the lowest levels in counterinsurgency. In just seconds, a young man must make a decision that may haunt or end his life' (Petraeus 2007: xxvii–xxviii). Ricks sums up the conventional protocol attached to operating roadblocks and the ethical responsibilities attached to following protocol:

Nor was checkpoint duty pleasant for soldiers: They were given three seconds in which to act against a suspicious vehicle, with the first shot fired into the pavement in front of the car, the second into the grille, and the third at the driver. 'We told them, you don't have the right not to shoot,' recalled Lt. Gen. John Sattler, a commander of the Marines in western Iraq. 'It's not about you. You are being trusted by everybody behind you. You are the single point of failure.'

(Ricks 2007: 361)

Sattler's understanding of checkpoints marks a key dimension of the ethical dilemma faced by troops in Iraq. If US combatants fail to stop a suspicious vehicle that is carrying a bomb, then some people – combatants or civilians – will face a mortal threat; if they fail in their duty there is a direct possibility that people will die. Yet by shooting at an approaching vehicle, combatants risk killing innocent civilians. This would seem to imply that a lingering ethical tension hangs over the operation of roadblocks. Indeed, the authors of the COIN manual identify this suspicion, stating that 'no rules can eliminate the underlying conflict' (Petraeus 2007: xxvii). Nevertheless, there is a belief, expressed in the COIN manual, that the tension between protecting innocent drivers who approach roadblocks and preventing threats from breaching the checkpoint can be reconciled: 'The manual seeks to present force protection and civilian protection as reconcilable' (Petraeus 2007: xxviii). As such, the COIN manual reinforces Walzer's overarching belief that conventions allow us to resolve the ethical questions raised by war. Let us now look at a few examples of roadblocks in action derived from a reading of first-hand accounts.

The first case we will look at is from Wright. Wright describes how Marines, during the Baghdad offensive, were charged with setting up a concertina wire roadblock outside the town of Al Hayy. A shooting incident occurred when a truck approached the blockade, travelling at 30–40 mph. The Marines fired a warning burst, followed by a second; at this point the truck's headlights pointed straight at the Marines' position, blinding them – to which the command is given to 'light it the fuck up'. The Marines shot at the truck until they killed the driver and it jack-knifed to a halt. Three men jumped from the truck attempting to run from the gunfire and were subsequently gunned down. The Marines immediately began to doubt their actions, feeling that they would carry the guilt home with them (Wright 2005: 278–280). The second example comes from another journalist, Chris Hondros. Hondros describes a night-time roadblock set up in occupied Baghdad by a foot patrol. A car approached the roadblock and soldiers fired warning shots; when the car sped up they fired into the car, killing a couple and injuring their six children to varying degrees. The patrol later discovered that the family was returning home from dinner with relatives, and were trying to get home before the imposed curfew. The family sped up as they perceived the gunfire as coming from behind them and merely wanted to escape (Hoyt and Palatella 2007: 159–160). Our final example comes from former US soldier Garrett Reppenhagen. Reppenhagen states that his unit set up a hasty checkpoint in order to be unpredictable and catch insurgents by surprise.

Reppenhagen provides a lucid insight into the mind of a combatant manning a checkpoint: 'You're thinking it could be a car bomb . . . You're thinking about the time somebody didn't fire and he got in trouble for not firing because they said he was endangering his unit. You're thinking about the guy that did fire another time and killed an innocent' (cited in Gutmann and Lutz 2010: 102). In this case, although the driver saw the checkpoint late, the soldiers refrained from directly shooting the driver long enough for him to stop. Nevertheless, they did feel 'justified to rip him out of his car and throw him on the ground and put him in handcuffs' (ibid.).

All three examples highlight why those operating the roadblock were ethically uncertain at the precise moment of their actions. They did not know if they were shooting at resistance fighters or frightened civilians. This uncertainty cannot be alleviated by the conventional procedure of warning shots, primarily because there is an implicit understanding – identified by many of the protagonists – that warning shots may not be interpreted as a call to stop the vehicle, and may in fact have the reverse effect, causing civilians to speed up in order to escape the gunfire. This is not to say that better conventions and operating procedures cannot be formulated; indeed they should. But what is equally important is acknowledging that uncertainty is rooted in the dynamics of the roadblock: a vehicle approaches the checkpoint, and a group of soldiers must determine the threat posed by this vehicle and subsequently act upon this threat. Even in cases where combatants have killed resistance fighters transporting car bombs, the structure of the act remains the same. It is only in retrospect that combatants can know if a potential bomber or civilian has been killed; it is only after the soldier decides to fire or refrain from firing that any concrete knowledge of the threat is possible. As Wright argues, combatants, by firing, engage in a game of moral chance and 'when it's over, he's as likely to go down as a hero or as a baby killer' (2005: 230–231). The play of uncertainty implicated in the operation of roadblocks highlights the undecidability evident in the determination of threat: the decision to fire or not to fire on the approaching vehicle is made in the absence of definitive knowledge, and the future will not produce more stable knowledge as the action has already conditioned the future context.

While the examples of roadblocks are not necessarily representative of all battlefield determinations of threat, they help elucidate a number of important points. Walzer's understanding of threat determination functions upon the assumption that those determining potential threats can do so objectively in the present. However, in Iraq, US troops were expected to determine threats when they were engulfed in a sense of personal danger. They were expected to accurately identify threats in an environment where every pile of trash or animal carcass was viewed suspiciously as a potential IED (Finkel 2011: 47), mobile phones were used to detonate roadside bombs and alert resistance fighters to US military presence (Filkins 2009: 187), and even children were employed as mortar spotters (Mejia 2008: 90–92). In what Wright calls a 'horrorscape of war', is it possible for combatants to pinpoint threat while operating under conditions of extreme duress? Wright summarises the physical and psychological impacts of combat thus:

In addition to the embarrassing loss of bodily control that 25 percent of all soldiers experience, other symptoms include time dilation, a sense of time slowing down or speeding up; vividness, a starkly heightened awareness of detail; random thoughts, the mind fixating on unimportant sequences; memory loss; and, of course, your basic feelings of sheer terror.

(Wright 2005: 182)

Under such conditions, combatants often claim to lapse into muscle memory, essentially acting through training and conditioning rather than conscious thought. As Mejia argues, within the midst of combat, complex moral analysis gives way to a mortal fear of dying (2008: 206). This fear is played out in Wright's narrative: at the start of a particular battle the Marines were adamant that they must avoid shooting at women and children; nevertheless, when they realised their own lives were in real danger, moral anguish gave way to a desperate survival instinct (2005: 127–130). This does not imply that the Marines were amoral or even unprofessional; in fact, what it indicates is that while combatants may be prepared to risk their lives, they will, nonetheless, still do everything in their power not to die. But this also has implications for the types of decisions made during war fighting. When a combatant is surrounded by the direct threat of death for a sustained period, a glint of sun or a camera flash can often be mistaken for a muzzle flash. In other words, the conditions of war produce contexts in which combatants' ability to objectively determine threat is further undermined. In this respect, US troops' determination of threat in Iraq is embroiled within the interplay between their daily exposure to mortal risk and the fear that this risk will be realised if they fail to act against potential threats. This tells us that the ethics of war are negotiated within fluid contexts, where adjudication of threat rests primarily in the hands and through the scopes of frightened young men and women. Ethical responsibility, in this sense, is enacted through a myriad of individual decisions, often dictated by impossible time constraints such as those evident in the case of roadblocks. As such, ethics is fundamentally infused with uncertainty and irresponsibility. Ethical responsibility cannot be resolved via objective determinations of threats, because determinations are produced within the specific contexts of war. Hence, the ideal of objective determinations of threat is unhelpful in explaining how combatants calculate risk in war and, therefore, how they determine who has forfeited their rights and why.

Freedom and sacrifice

The critique of Walzer's justifications for the forfeiture of combatant rights highlights the inadequacy of the War Convention in resolving the ethical problems associated with war. Walzer fails to justify the loss of rights upon any clear conception of individual action in war. Instead, his argument is premised upon the belief that life and liberty are simply not applicable to the role of combatant. Ultimately, Walzer points to a singular justification for the revocation of combatant rights: *combatants lose their rights because they have chosen to become combatants.* Walzer depicts combatants as fundamentally different to all other people; in Hedley Bull's terms, he refuses to conceive of them simply as human (1979: 593). It is only through the assumed de facto threat posed by combatants that Walzer can justify their slaughter as wholly moral. It is not the individual actions of the soldier that constitutes a threat worthy of the forfeiture of rights; it is because the soldier is a soldier that their threat is illimitable on the battlefield. Walzer, in turn, unequivocally endorses the killing of combatants in war on the grounds of their role rather than their actions:

> Soldiers as a class are set apart from the world of peaceful activity; they are trained to fight, provided with weapons, required to fight on command. No doubt they do not always fight; nor is war their personal enterprise. But it is the enterprise of their class, and this fact radically distinguishes the soldier from the civilians he leaves behind.
>
> (Walzer 2006a: 144)

Because it is the combatants' profession that sets them apart from the innocent and places them into a position where they fight in the first instance, it is crucial that we investigate how a civilian becomes a combatant, thereby forfeiting their rights.

Given the importance attached to the act of becoming a soldier, and the centrality of the proposition that the combatant '*allowed* himself to be made into a dangerous man', it is surprising that Walzer begins his discussion on war by asserting that soldiers do not fight freely. Walzer argues that combatants only fight to defend the safety of their community: 'he has to fight (he has been "put to it"): it is his duty and not a free choice' (2006a: 27). The claim that the decision to fight is 'not a free choice' is interesting, because it suggests that the act of becoming a combatant is not necessarily the decision of individual combatants. Further to this, Walzer contends that the decision to fight is not freely taken in cases where enlistment is voluntary (2006a: 28). In fact, he even maintains that mercenaries do not fight freely if they fight due to economic necessity (2006a: 27). Walzer presents us with a depiction of war fighting in which those who do the fighting – and therefore lose their title to life and liberty – are not fastened to the role of combatant by their own choosing. He sums up the tyranny in stark terms:

> Hence the peculiar horror of war: it is a social practise in which force is used by and against men as loyal or constrained members of states and *not as individuals who chose their own enterprises and activities.*
>
> (Walzer 2006a: 30, italics mine)

Paying particular attention to Walzer's choice of terms, let us reflect upon how civilians become active combatants. In Walzer's argument, a free citizen loses their rights through the act of becoming a soldier and, thereby, becoming a dangerous man. However, this act is, in Walzer's view, not a free choice; it is not an activity of the soldier's own choosing and, therefore, its integrity as an act justifying the loss of a combatant's rights is compromised. If we recall Walzer's two absolute rights, life and liberty, surely the forced enlistment of soldiers, whether by moral obligation or legal duty,[23] constitutes a breach of the latter right. Walzer's conception of forced fighting has some resonance with the reasons that US combatants give for enlisting. For example, many recruits argue that they joined through economic necessity or as a means to pay for university education (Gutmann and Lutz 2010: 21, 29, 72).[24] Nevertheless, soldiers enlist for numerous other reasons: to face the unknown (Wright 2005: 51); to test themselves against adversity (Hennessy 2010: 32); and to protect the US's core principles (Fick: 2007: 5). However, what is clear from many soldiers' experiences in Iraq is that they would rather be at home in the US than risking their lives in a foreign land; as a US soldier confesses: 'They say on TV that the soldiers want to be here? . . . ain't nobody wants to be here' (cited in Finkel 2011: 117). What is perhaps more important, in regard to this discussion, is that Walzer premises his conception of morality in war on the assumption that combatants would prefer not to fight: 'soldiers would almost certainly be nonparticipants if they could' (2006a: 30). In other words, Walzer begins from the presumption that combatants would choose not to fight if they were given a free decision.

In this sense, Walzer's justification for the killing of combatants hinges upon a prior violation of combatants' right to liberty, for which no justification is given. Prior to enlistment, combatants are non-combatants, innocent and immune from attack. It is only when they are forced to become dangerous men that they are transformed into legitimate targets. Mejia sums up the hopeless frustrations of soldiers in Iraq by arguing, '. . . my misfortune was tied to a decision I had made at nineteen when I signed a military contract and forfeited most of my rights' (2008: 134). Indeed, if a US soldier refuses to deploy, it is highly likely that they will still lose their right to liberty via court martial, especially considering US military trials boast extraordinarily high conviction rates; these were, for example, 93.7 per cent in 2010.[25] The other alternative for troops is to flee from the US itself, sacrificing the life they had built there and their citizenship status. In Walzer's terms, they would need to sacrifice their maximal life and communal membership; therefore, as soon as an individual signs a military contract, their right to life and liberty is compromised in some crucial ways. Importantly, the perceived loss of rights, in Walzer's argument, is not a result of the actions of individual combatants. Rather, rights are lost because of the actions of the combatant's own state or its adversary in starting the war – Walzer's conception of *ad bellum* aggression. Combatants do not choose to start the war and they do not freely choose to fight in it. According to Walzer, combatants are 'coerced moral agents' and 'men whose acts are not entirely their own' (2006a: 306, 309). Walzer cannot reverse his contention that soldiers do not fight freely, for to do so would implicate combatants in the justness of their cause, thus eliminating the moral equality of soldiers. It is only because soldiers do not fight freely that war is not their crime; and if this condition were to be reversed we could only justify killing in the name of just ends. Walzer clearly does not want, or intend, to make this argument because it would create 'a new class of generally inadmissible acts and of quasi-rights, subject to piecemeal erosion by soldiers whose cause is just – or by soldiers who believe that their cause is just' (2006a: 230). The moral equality of soldiers is imperative to Walzer's theory, to his separation of *jus ad bellum* and *jus in bello*, and to the foundation of the War Convention. Yet this imperative is founded upon the unjustified revocation of combatants' right to liberty. Two incompatible imperatives clash in the foundation of Walzer's rules of war: combatants can be attacked because they allowed themselves to become dangerous men; but they do not fight freely and would choose not to fight if they could. Rights are not forfeited through identifiable acts of individual combatants and, as such, ethical sacrifice remains rooted in the killing of combatants. Combatants' title to life and liberty is sacrificed in the name of the ideal of communal self-determination.

Conclusion

As we have discussed, in Walzer's argument, rights are not withdrawn from individual people; they are withdrawn from the role of combatant. In this sense, a role that individuals do not freely take upon themselves becomes the defining attribute of their personhood. In Derridean terms, the role of combatant signifies a modern conception of *pharmakoi*. Derrida outlines the role of *pharmakoi* in ancient Athens (1981a: 128–134). *Pharmakoi* were foreigners viewed as 'sub-human' individuals by the Athenians and who were taken from outside Athens and housed in the heart of the city bounds. These foreign prisoners were kept in the city for the specific purpose of sacrifice in the event of disaster, drought or famine. In turn, the sacrifice was ritually undertaken outside the city grounds. The sacrifice of the *pharmakoi* signified the symbolic removal of the malignancy falsely depicted as the cause of the disaster. *Pharmakoi* were taken into the city so they could be identified as the cause of future tragedies and expelled to demonstrate the rulers' ability to protect the city from further disasters. They occupied the role of sacrificial scapegoats whose sacrifice reassured the safety, sanctity and purity of the city. In this way, *pharmakoi* symbolised a constitutive outside that was ritually sacrificed to preserve the idea of inner purity and security.

In terms of Walzer's argument, combatants represent an interesting iteration of *pharmakoi* in contemporary US society. Although combatants are primarily US citizens and are already part of the US community,[26] rather than foreigners taken in from the outside, Walzer casts them outside the sanctity of rights: when war is declared, and the community faces a perceived threat, the combatant-*pharmakoi* forfeits their rights. Walzer believes that this forfeiture is crucial in order to ensure the safety of the community and its self-determination, and the broader regime of minimal rights. The combatant must be reduced to an excommunicated outside so the inside can be protected. Importantly, by withdrawing rights from the combatant, Walzer claims that it is possible to conduct war without sacrificing ethical responsibility. Combatants are placed outside the protection of rights as a means to render violence morally unproblematic; however, this revocation of combatant rights cannot be sustained in the way Walzer wants. His central argument is that rights can only be surrendered on the basis of an individual's actions; yet the forfeiture of combatant rights is predicated solely upon assumptions about the role of combatant.

Forfeiture is underpinned by the belief that combatants are dangerous and will fight on command. Nevertheless, Walzer simultaneously contends that combatants do not freely choose to become combatants; they do not fight freely and their actions are not fully of their own choosing. The justification of the killing of combatants is, therefore, premised upon the unjustified revocation of combatants' absolute right to liberty. In Walzer's system, combatants lose their rights due to decisions made about, but not by, them; and in this way, the defence of community is made possible through the dereliction of duty to individual combatants. War is justified via the unjustified sacrifice of combatants' absolute right to life and liberty.

Notes

1 It is interesting to note the masculine connotations of war fighting in Walzer's analysis. Combatants are almost uniformly denoted in masculine terms, while the 'innocent' in war are often described as women and children.

2 The next chapter will directly address Walzer's justifications for the unintentional killing of non-combatants in war.

3 See *Just and Unjust Wars*, chapter 11.

4 I will address the implications of the claim that combatants do not fight freely at a later point in the chapter.

5 See *Just and Unjust Wars*, chapter 3, pp. 33–47.

6 These two groups are conjoined in *jus post bellum*: politicians must ensure that there is a clear plan for post-war justice and combatants must ensure, to the best of their ability, that this plan is achieved.

7 For an alternative critique of the concept of moral equality, see Jeff McMahan (2009), *Killing in War* (Oxford: Oxford University Press).

8 This is possibly a result of the schooling that Marine Corps officers receive in just war theory.

9 For example, every US soldier deployed to Iraq was required to settle their estate, compile a will and provide a contingency plan in the event of their death. They are even required to pose for a photographic portrait in front of the US flag that can be displayed during their funeral and supplied to the media if they are killed in active duty (Finkel 2011: 11–12).

10 Among the most interesting aspects of the symbolic acceptance of death in Iraq were the so-called ranger graves used by US troops to protect themselves against stray shrapnel. Wright describes how Marines were required to dig these graves every time they were granted rest. The symbolism involved in the act of digging and sleeping in one's own grave was not lost on the Marines in Wright's account (2005: 92).

11 In fact, one of the main reasons Walzer refuses to hold individual combatants accountable for the crime of aggression is that they are so often duped by their leaders into believing that amoral wars are fought for just reasons.

12 In turn, the Iraqi resistance held similar views of US troops (Shadid 2006; Cockburn 2007).

13 This idea is central to McMahan's (2011) critique of Walzer. McMahan contends that unjust combatants have no right to kill because by allowing them to kill we are enhancing an unjust cause. However, it must be noted that McMahan believes that the justice of a particular war can be definitively deduced from the onset. As such, McMahan operates within a system in which the ethics of war can be resolved via universal rules and norms.

14 For a broader discussion on the valuation of lives in war, see Judith Butler (2010), *Frames of War: When is Life Grievable?* (London: Verso).

15 This implication is particularly pertinent in regard to the Iraq War because the US Military has implemented a widespread policy of 'Stop-Loss'. This policy allows the US Military to extend a service person's contract without their consent. A congressional report stated that 185,000 troops serving in Iraq and Afghanistan (between 2001 and 2009) have been subjected to Stop-Loss (Olsen 2011: 426).

16 Figure taken from US Department of Defense website, http://www.defense.gov/news/casualty.pdf

17 Gutmann and Lutz stress that the RAND study is a much lower estimate in contrast to a Veterans Affairs study that found between 30 and 40 per cent of troops returned with some form of psychological trauma.

18 For example, as of May 2006, 1,600 US children had lost a parent due to the wars in Iraq and Afghanistan (Chartrand and Siegal 2007: 1).

19 Walzer also equates this to the question of a soldier's ability to bear arms: 'That right (to immunity) is lost by those who bear arms "effectively" because they pose a danger to other people' (2006a: 145). He nonetheless does not explain what 'effectively' means in his argument, or how we can judge the effectiveness of an individual combatant's capacity to use their weapons.

20 Ricks points to the infamous 2003 car bomb attack on UN headquarters that resulted in the mass withdrawal of UN personnel from Iraq, thereby illustrating the strategic necessity of roadblocks.

21 It must also be noted that because many roadblocks took place at night and some roadblocks were camouflaged to reduce the risks to the troops operating them (Hoyt and Palatella 2007: 159–160), it was often unclear whether motorists approaching roadblocks were actually aware that they were entering a restricted area.

22 In some instances, smoke grenades have been used, but this can potentially decrease the chances of the driver stopping because the grenades blind their line of vision (Wright 2005: 350).

23 Walzer argues that democracies have an increased coercive power in enticing citizens to enlist (2006a: 35).

24 There are major socio-economic and cultural dimensions to Iraq and Afghanistan's so-called economic draft that Walzer's theory does not address. However, if justice is the aim of Walzer's theory, then some attention must be paid to social and cultural cleavages from which the modern US Military is primarily drawn, and to the sectors of society who largely avoid military service. See Olsen (2011).

25 See the 'Annual Report Submitted to the Committees on Armed Service United States Senate and the United States House of Representatives and to the Secretary of Defense, Secretary of Homeland Security, and Secretaries of the Army, Navy, and Air Force Pursuant to the Uniform Code of Military Justice For the period October 1, 2009 to September 30, 2010', http://www.armfor.uscourts.gov/newcaaf/annual/FY10AnnualReport.pdf

26 The cultural composition of the US Military is interesting here because military service is still employed as a citizenship pathway: for example, see Mejia (2008). Olsen's (2011) analysis also points toward an understanding of military service, within US society, as a means to achieve legitimate citizenship. Olsen argues that minorities, primarily first and second generation emigrants, on the margins of US society are actively encouraged to join the military as a means to prove their loyalty to their 'new' homeland. For example, Olsen argues that the disproportionate number of US citizens of Hispanic origins that served in the Iraq War highlights the desire of Hispanic communities to prove that they are 'real Americans'.

5 Double effect and its parasites

Introduction

The intentional killing of civilians is, for Walzer, one of the most morally reprehensible acts that can be committed in war, and is intimately linked to unjustifiable terrorism (2005: 51–66). Walzer outlines what he believes to be the crucial distinction between just war and terrorism; it is the moral difference 'between aiming at particular people because of things that they have done or are doing, and aiming at whole groups of people, indiscriminately, because of who they are' (2006a: 200). In other words, just warriors target identifiable combatants who have forfeited their rights, while terrorists target civilians, thereby violating absolute rights. In this sense, the possibility of fighting a just war hinges upon the refusal to target people who have not forfeited their rights. Despite Walzer's opposition to the killing of civilians, he recognises that avoiding civilian casualties is a practical impossibility in modern war. Walzer argues that damage to civilians and civilian property – what is commonly referred to by the term collateral damage – is an unavoidable reality of warfare. To reconcile this inevitability he attempts to justify collateral damage through the traditional just war doctrine of double effect:

> Soldiers could probably not fight at all, except in the desert and at sea, without endangering nearby civilians . . . Double effect is a way of reconciling the absolute prohibition against attacking non-combatants with the legitimate conduct of military activity.
>
> (Walzer 2006a: 153)

In short, double effect renders the dangers necessarily imposed on civilians morally justifiable; as such, it is the second key foundation of the War Convention. While the principle of non-combatant immunity attempts to justify the killing of combatants, the doctrine of double effect explains why the risks imposed on civilians during war do not necessarily compromise the possibility of fighting in an ethically responsible way. These dual foundations combine to produce an account of warfare in which putting both combatants and non-combatants in danger is not tantamount to a violation of rights.

Walzer distinguishes acceptable collateral damage from terrorism by high-lighting the difference in the guiding intention. On the one hand, just fighters aim at legitimate targets and sometimes *unintentionally* harm civilians; while on the other hand, terrorists *intentionally* aim to kill civilians. Intention, in this way, becomes the key moral factor in Walzer's separation of just war from terrorism. The argument in the last chapter questioned Walzer's justification of the forfeiture of combatant rights; this chapter, in turn, provides a critical examination of his justification of civilian causalities. To this end, I focus upon the doctrine of double effect and its justifications. I argue that double effect fails to justify collateral damage, and that intention cannot provide an adequate means for justifying violence; in addition, I contend that the doctrine has much broader implications than traditionally conceived. Double effect is primarily associated with the immediate effects of military targeting, bombing, shelling, and so on. This chapter, however, points toward a more expansive conception. Drawing upon the depiction of ethics as response, this chapter illustrates why the unintended and unforeseen effects of war fighting are constitutive of the terrain in which future ethical relationships take place. As such, it draws attention to the ways unintended consequences of warfare produce new contexts and, therefore, new possibilities for ethical relationships.

The doctrine of double effect

Walzer outlines four primary conditions of the classical just war conceptualisation of double effect:

1   It [the military operation] is a legitimate act of war.
2   The direct effect is morally acceptable.
3   The *intention* of the actor is *good*, that is, *he aims only at the acceptable effect.*
4   The good effect is sufficiently good to compensate for allowing the evil effect.
                                                            (Walzer 2006a: 153, italics mine)

While all four conditions must be met, Walzer argues that the *intention* of the protagonist is the deciding moral factor. In a similar manner to his critique of terrorism, he maintains that the intention not to target civilians is central to fighting justly: 'The burden of the argument is carried by the third clause . . . the killing of soldiers and nearby civilians, are to be defended only insofar as they are the product of a single intention, directed at the first and not the second' (2006a: 153). In short, intending to aim solely at the legitimate target absolves the combatant of all responsibility for any negative unintentional consequences.

The COIN manual echoes Walzer's sentiments by stating that 'Soldiers and Marines may take actions where they knowingly risk, but do not *intend*, harm to non-combatants' (Petraeus 2007: 245, italics mine). US troops in Iraq were thus permitted, in certain instances, to risk civilian lives provided they did not directly intend to harm them. Wright (2005) offers an illustrative example of this principle in action. He describes how artillery was deployed to subdue the hostile town of Nasiriyah and the 3,000 to 5,000 Saddam loyalists opposing the US advance through the town:

> For some reason reporters and antiwar groups concerned about collateral damage in war seldom pay much attention to artillery . . . But the fact is, the Marines rely much more on artillery bombardment than on aircraft dropping precision-guided munitions. During our thirty-six hours outside Nasiriyah they have already lobbed an estimated 2,000 rounds into the city. The impact of this shelling on its 400,000 residents must be devastating . . . I feel relief every time I see another round burning through the sky. Each one, I imagine, ups the odds of surviving.
>
> (Wright 2005: 152–153)

Wright captures the contradictory horror and relief of the artillery strike: troops knew that damage was most likely being inflicted on civilians but simultaneously realised that this damage could be the difference between them living or dying. Yet Wright's discussion puts intentionality in a rather suspect position. If those professing good intention – i.e. to aim solely at legitimate targets – potentially increase their own safety by endangering civilians, could they not simply feign good intentions? Could militaries not, for example, intentionally risk innocent lives in order to protect their own troops as Wright's discussion implies? What is important, in the context of Walzer's argument, is that intentions are ambiguous; we are not certain that the goodness of the act is truly intended rather than fallaciously professed.

Derrida explains that deceitful mimicry and simulation is always possible where questions of responsibility are concerned (2009: 27). In terms of double effect, it is always possible to feign good intention in order to advance a military objective through the killing of civilians. For example, it is always possible for combatants to intentionally risk civilian lives and subsequently claim that they did not realise any civilians were at risk. The conventional understanding of double effect thus opens the possibility of infelicity and dishonesty. For this reason, Walzer argues that the doctrine needs a supplementary[1] condition to adjudicate over the alleged *goodness* of the intention: the concept of *due care*. Walzer explains that due care balances combatants' acceptance of risk against the risks imposed on civilians: 'The intention of the actor is good, that is, he aims narrowly at the acceptable effect: the evil effect is not one of his ends, nor is it a means to his ends, and, aware of the evil involved, he seeks to minimize it, accepting costs to himself' (2006a: 155). In this way an economy of risk is installed to temper the possibility of intentional infelicity. Militaries that refuse to reduce the risks of collateral damage by accepting the costs fail the test of good intention and their acts are unjustifiable. This idea is reaffirmed in the COIN manual, which encourages combatants to 'preserve non-combatant lives by limiting the damage they do' and 'assume additional risk to minimize potential harm' (Petraeus 2007: 247). In Walzer's argument the acceptance of risk acts as a form of insurance against infidelity of intention – what I will term 'risk-as-insurance'. Combatants' acceptance of risk assures us that they do not intend to harm civilians. Nevertheless, Walzer recognises that this conception of risk-as-insurance is complicated by the tension between the strategic necessity of winning the war and the moral necessity of protecting civilians; that is, if combatants take on too many additional risks in their attempt to protect civilians, they risk losing the war. Walzer argues, then, that due care is an attempt to balance these dual requirements: soldiers must accept extra risks to minimise potential harm to civilians, but they are not required to undertake risks that place legitimate military operations in danger of failure. 'War,' he maintains, 'necessarily places civilians in danger; that is another aspect of its hellishness. We can only ask soldiers to minimise the dangers they impose' (2006a: 156). Due care, in this respect, constitutes a negotiation of two contradictory and heterogeneous necessities: the necessity to protect civilians butts up against the necessity to win the war. Winning a war, therefore, necessitates the *unintentional* sacrifice of civilian lives.

Walzer contends that the War Convention invites soldiers 'to calculate costs and benefits only up to a point, and at that point it establishes a set of clear cut rules' (2006a: 131). We should, therefore, expect the War Convention to provide clear rules of double effect and due care. Walzer responds by stating that the limits of due care are fixed at the point where undertaking further risks would doom the military venture or make subsequent military actions impossible (2006a: 157). Walzer, in this way, depicts a rather idealised conception of risk calculation: the image of a soldier marching toward their target knowing that each step decreases the risk they pose to civilians, but also aware that each step increases the risk that the attack will fail. In this idealised narrative, the soldier stops at the precise line between success and failure, and launches the attack. Good intention is validated by the acceptance of risk without fundamentally compromising the strategic aim. Nevertheless, because intentionality – determined by risk-as-insurance of fidelity – is valorised as the governing centre of morally justified collateral damage, we must ask how this ideal point is determined in the midst of war. How do we really know whether the condition of due care has been met in an actual wartime situation?

Pardon me for not meaning to . . .

Walzer's conceptualisation of due care underscores a key aspect of his understanding of moral judgement in war. The specific purpose of due care is to demonstrate the goodness of intention not to harm civilians. Without due care there would be no way to prove the alleged goodness of the intention and, therefore, the justness of the act; for due care is formulated as a means to communicate and thereby authenticate the justness of military actions that endanger civilians. In important respects, the possibility of justifiable war presupposes that acts of war are communicable. Walzer's theory implies that the justness of the cause and the justness of the way the war is fought can be communicated to ordinary people. If the justness of war could not be communicated there would be no possibility for moral judgement in the ways that he suggests – i.e. if it is possible to regard a war, or an act of war, as just or unjust, this implies that wars and their operations must remain communicable. However, if we recall the discussion on Derrida's understanding of communication and iterability, Walzer's understanding of how intention is communicated is called into question. Walzer's formulation of double effect presupposes a particular understanding of intentionality and communication. Primarily, he assumes that the good intention can be fully actualised on the battlefield and fully communicated to those judging the justness of the act. Although Walzer proposes due care as a means to test the authenticity of intention, this still relies on a telos of pure fulfilment. He presents us with a homogenous movement in which intentionality is fully translated into action: the combatant aims at the legitimate target with the singular intention of hitting this target and this target alone, and they undertake personal risks in order to reduce the negative effects to civilians. In Walzer's terms, the action must be the result of *a single good intention*. It is only by maintaining the purity of the fulfilled good intention, uncorrupted by the negative unintended consequences, that Walzer can exempt the deaths resultant from collateral damage from the absolute prohibition on the killing of civilians. Intention is, in this way, split from the unintended consequences; and the means rather than the ends determine the moral signature of the act. In Walzer's model, once the good intention has been authenticated through a process of due care, it stands fulfilled in its entirety; and the unintended residuum is viewed as structurally external to the intentional act. In other words, unintentional negative consequences, although lamentable, do not compromise the goodness and singularity of the intention; the good intention remains wholly good and wholly fulfilled despite the regrettable unintended effects.

The Derridean understanding of iterability challenges the possibility of fulfilled intention implied in Walzer's argument. As outlined in the third chapter, Derrida describes iterability as the ability of communication to function in the absence of its intended meaning. Iterability means that every communicable message must be readable and repeatable in the absence of the author and the singular intention of the message's production (1988: 8). In this respect, iterability signifies the becoming otherofintentionthroughcommunication–theriskofmeaningtosaysomethingother than intended. Iterability, however, is simultaneously a necessary component of communication. Without the possibility of being detached from its intended meaning, the communicated message could not possibly be read and repeated by another (or indeed by the self at another time). Derrida contends that

> [t]he unit of the signifying form only constitutes itself by virtue of its iterability, by the possibility of its being repeated in the absence of not only its 'referent,' which is self-evident, but in the absence of a determinate signified or *the intention of actual signification, as well as of all intention of present communication.*

> (Derrida 1988: 11, italics mine)

Because acts of communication must be able to function in the absence of the initial intention of their production, they are necessarily cut off from the singular intention present at the moment of their creation. In Derrida's words, they are 'divided and deported in advance' – and this is not an accident (1988: 56). What Derrida means by this is that the singular intention guiding an act of communication does not fix meaning. This is because acts of communication are addressed to other people who must interpret the meaning and, therefore, risk altering the intended meaning. Again, this highlights the alterity implicit in communication and, as I will explain, in every possible intentional action. Recognising the positive condition of iterability as a structural requirement for communication also constitutes the necessary negative limit of intentionality: iterability allows us to communicate an intended meaning while simultaneously denying the uninterrupted and uncorrupted transportation of the singular intended meaning. In the context of Walzer's argument, the negative limit of iterability challenges the pure fulfilment of good intentions implied in the doctrine of double effect.

A brief discussion on signatures will help clarify the relationship between intention and communication. Like Walzer's depiction of due care, signatures function as a seal of authenticity and fidelity; however, signatures also illustrate why iterability limits the fulfilment of intentionality. Derrida argues that the signature must be the product of a singular intention, i.e. to authorise a singular event in the name of a singular author whose future absence is inscribed in the signature's very production. The signature is often employed, for instance, to assure us that the named signer endorses a text to which their name is attached. Nevertheless, in order to authorise on behalf of the signer, the signature must be repeatable and imitable; a signature must be iterable. Therefore, the signature must detach itself from the intention guiding its production in order to function. As Derrida maintains, it carries a divided seal (1988: 20). The signature must be iterable in order to function as a mark that can be read and thereafter authenticated by a third party; yet because signatures are iterable (readable, repeatable), they must necessarily run the risk of being utilised for an event distinct from the intention of their production. For example, the signature runs the risk of being defrauded, of being attached to a text or event that the signer does not wish to endorse. The divided seal of the signature is both a necessary component of its operational structure and the impossibility of its rigorous purity; its positive condition and its negative limit. In the structure of signatures we see another example of the impossibility of fulfilment of self-presence: the possibility of the becoming other of intention implicit in every conceivable communicable act undoes the completion of fully actualised intentionality. In terms of due care, this implies that Walzer's purported seal of authenticity can, too, be distorted for aims other than those he intends.

Walzer's formulation of double effect cuts the singular intention off at the point at which it is actualised. The specific action is de facto just or unjust at the precise moment it is executed, and is judged on the basis of due care. Consequences are removed from the moral equation because the good intention has already been assured via due care prior to the production of any consequences. The act is judged to be just or unjust before its effects have occurred.

Walzer's conception of intentionality, in this respect, echoes advice outlined to US Marines in Iraq by their commanders: 'It doesn't matter if later on we find out you wiped out a family of unarmed civilians. All we are accountable for are the facts as they appeared to us at the time' (cited in Wright 2005: 53). In this sense, Walzer's model and the Marines' advice justify actions based on professed intentions and not actual end results. By contrast, the Derridean understanding of intention suggests that the risk of negative unintended consequences is a structural fixity in any act of war and, as such, cannot be discounted from the moral signature. Derrida's critique of this conception of intentionality does not, however, constitute a denial of the role of intention; it simply stresses the need for a new typology in which intention has a place but cannot govern the entire system (1988: 18). In short, Derrida's argument is primarily questioning the purported *undividedness* of intention (1988: 105).

The key lesson from Derrida's typology is that unintentionality is always already at work from within every possible intentional movement: '. . . it leaves us with no choice but to mean (to say) something that is (already, always, also) other than what we mean (to say)' (1988: 62). In terms of double effect, it leaves us no option but to *mean* to risk unintentional negative effects: as soon as a combatant fires at any target they necessarily risk adversely affecting civilian lives. The risks posed to civilians, therefore, cannot be logically or cognitively detached from the intention to launch the attack. In Derrida's words,

> [a]s soon as [*aussi sec*] a possibility is essential and necessary . . . it can no longer either de facto or de jure, be bracketed, excluded, shunted aside, even temporarily, on allegedly methodological grounds. Inasmuch as it is essential and structural, this possibility is always at work making *all the facts*, all the events, even those which appear to disguise it.
>
> (Derrida 1988: 48, original italics)

Derrida reminds us that once a possibility is structurally necessary it cannot be excluded as a risk that is accidental and exterior. Because military actions, according to Walzer, necessarily put civilians in danger, the risk to civilian life cannot be discounted as some ditch which intention can fall into unintentionally. Unintended consequences are never simply external because they co-found the very roots of the intentional act. Ultimately, an act of war could not possibly constitute an instance of *double* effect if the risk of negative effects were not already implicated in the act's inception. The word 'double' already presupposes the potential unintended consequences. In this sense, the ideal structure of 'good intention' is intricately dependent upon the necessary possibility of the negative *unintended* consequences. In other words, a successful intentional act is only possible if that intention can fail. Military actions risk negative effects because this risk is a structural necessity in any conceivable action. Therefore, in the absence of fully actualised intentionality, we are left only with the relative purity of intentional acts, which are judged in relation to each other and not according to an illusionary ideal of purity (Derrida 1988: 18). Double effect, therefore, cannot maintain the purity of the singular good intention in its entirety. Instead, it is judged on the basis of the commitment to minimise negative civilian impacts.

Nevertheless, Walzer clings to a typology that valorises the possibility of a fully realised intention detached and separate from the necessary risk of its failure:

> A soldier must take careful aim *at* his military target and *away from* non-military targets. He can only shoot if he has a reasonably clear shot; he can only attack if direct attack is possible. He can risk incidental deaths, but he cannot kill civilians simply because he finds them between himself and his enemies.
>
> (Walzer 2006a: 174, original italics)

Again, paying close attention to Walzer's language, we can see that the singular good intention (to hit the military target) is uncorrupted by the spectre of the unintentional (the risk of incidental deaths). Walzer, however, clearly asserts that the principle of double effect can only be employed in cases where the risk of unintended civilian deaths is directly evident: 'Officers can only speak in its terms, knowingly or unknowingly,[2] whenever the activity they are planning is *likely* to injure non-combatants' (2006a: 152–153, italics mine). In this sense, Walzer's singular good intention is predicated upon the risk of unintended effects. In fact, this risk is not simply a distant possibility, but on the very cusp of actualisation, because Walzer maintains that militaries only talk in terms of double effect when their actions are *likely* to injure combatants. During the 2003 invasion of Iraq, for example, Donald Rumsfeld was required to approve any strike likely to kill thirty or more civilians. Rumsfeld duly approved all of the fifty submissions that were made (Gregory 2004: 207). If the principle of double effect is only called upon in instances in which military actions are likely to injure civilians, then the risk of civilian harm must be intended if the target is fired upon, despite this knowledge. In Walzer's terms, combatants who aim at legitimate targets, under the principle of double effect, must not only intend the possible risk of negative effects, but also their *likely* realisation. Thus, Walzer leaves us with an entirely pragmatic concept of intentionality: the combatant is not judged according to the actualised fulfilment of their singular good intention, but against the relative purity of a divided intention. The combatant is judged on the basis of the extent to which the 'good' intention is achieved and the 'bad' intention (the likely risk) is minimised, and this judgement hinges upon the personal risk undertaken by the combatant to reduce the negative effects. In turn, this calculation becomes the defining moral aspect. As Asad argues, while a terrorist's conscience is never important, the sincerity of a military commander's intentions may be the crucial difference between an unfortunate necessity and a war crime (2007: 26). This transforms the supplementary concept of due care – the fidelity of intention proved by the risk accepted – into the primary adjudicator of the justness of acts of double effect.

In all good faith

By constructing a pragmatically defined conception of double effect, Walzer's justification for collateral damage rests upon 'the seriousness of the intention to avoid harming civilians, and that is best measured by the acceptance of risk' (2005: 137). Risk minimisation, in this way, becomes the determining factor in judging the morality of acts of war likely to injure non-combatants. It is, therefore, important that we analyse the implications of this for Walzer's overall theory of *in bello*. Walzer, unsurprisingly, is firm in his belief that risk-as-insurance acts as a barrier to imitations and insincere appeals to the doctrine of double effect. This barrier is of fundamental importance to Walzer's argument because he acknowledges that statesmen will tell lies in order to frame immoral activities undertaken in the name of military strategy in a moral way (2006a: 19). Walzer, then, anticipates that leaders will attempt to bend double effect to their strategic needs and, as such, double effect requires an enforceable system of due care through which appeals can be judged. Yet, in a simultaneous movement, the measures necessary to judge due care also signify the terminal breakdown of Walzer's theoretical fiction of a single undivided good intention. When discussing the topic of risk minimisation, he states that military strategists 'must take positive steps to limit even unintended civilian deaths (and they *must make sure* that the numbers killed are not disproportionate to the military benefits they expect)' (Walzer 2006a: 317, italics mine).[3] Now Walzer has worked his argument firmly into a paradoxical position. On the one hand, if military strategists are able to *make sure* that the number of civilians killed is outweighed by the expected military benefits, they have strategically planned for these deaths. Civilian deaths are explicitly intended (albeit as regrettable consequences of a necessary action) and, therefore, non-combatants have become the object of a military attack. Civilian deaths have, in Kantian terms, become a necessary means toward a military end because the attacking force has planned for the expected deaths of a specific number of civilians in their efforts to hit a legitimate target. On the other hand, if military strategists are unsure about the costs and benefits of the action, double effect can only ever be judged retrospectively on the basis of its consequences. Retrospective judgements are problematic for Walzer because they mirror realist ends–means justifications that he unequivocally dismisses as fundamentally amoral.[4]

Walzer's predicament is this: he can either present a model of double effect in which risk calculation is possible, hence condoning the intentional killing of a set number of civilians as a necessary cost of war fighting; or he can disallow calculation, thereby risking the transformation of double effect into a wanton realist excuse for the killing of civilians in the name of strategic necessity. Ultimately, Walzer opts for the former depiction, reluctantly conceding (by way of a footnote) that the due care component of double effect is partially a utilitarian argument: 'Since judgements of "due care" involve calculations of relative value, urgency, and so on, it has to be said that utilitarian arguments and rights arguments (relative at least to indirect effects) are not wholly distinct' (2006a: 156). Drawing upon this utilitarian argument, however, presents a major problem for Walzer's overall theory. Primarily, the utilitarian argument further undermines the idea of a singular undivided good intention. Because the calculations required by due care have already explicitly split intentionality (the intention to hit the military target and the intention to avoid killing more than X number of civilians), the rights component of double effect collapses. The act of double effect is not conditioned by a singular good intention (to hit the legitimate target), but by a divided intention to hit the legitimate target and avoid killing a disproportionate number of civilians in the process. Rights, in this way, are entered into an economy in which civilian lives are proportionate only to the value of the target. What we are left with is a manifest rule utilitarian argument: I can intentionally risk the deaths of a certain amount of innocent people provided the benefits derived from hitting the target outweigh the costs. In the absence of justification via singular good intention, the principle of double effect is supplanted by the concept of due care. The act is not validated by the intention not to target civilians but by a risk calculation that requires strategic planning for civilian deaths. The appeal to utilitarian calculation is important because Walzer assures us that his rights-based model of morality 'rules out calculation and establishes hard and fast standards' (2006a: 304). The doctrine of double effect, in other words, does exactly what Walzer intends his system of morality to block.

Walzer wants to avoid utilitarian rules for a number of important reasons. In a direct attack on utilitarianism he firmly asserts that:

> Utilitarianism, which is supposed to be the most precise and hard headed of moral arguments, turns out to be the most speculative and arbitrary . . . We have no unit of measurement and we have no common or uniform scale . . . Commonly what we are calculating is *our* benefit (which we exaggerate) and *their* costs (which we minimise or disregard entirely).
>
> (Walzer 2005: 38–39, original italics)

In Walzer's own terms, then, we should expect double effect to be employed in such a way that exaggerates the benefits to *our* military operation and minimises the costs to *their* civilians. We should expect the doctrine of double effect to be employed as a means to justify military strategy rather than protect moral rules. It is also important to remember that we are discussing the absolute right to life. Walzer steadfastly maintains that the civilians placed in danger during war have not forfeited their rights and, as such, for double effect to be justified, acts covered by the doctrine must not violate rights. Nevertheless, Walzer's depiction of due care suggests that risks to civilian lives must be balanced against the strategic value of the target, i.e. we must apportion a value to civilian lives. How, then, can we balance the absolute right to life against a military objective? In the case of utilitarianism, Walzer is quick to assert that such calculations are to be considered a form of bizarre accountancy: '. . . their inventions are somehow put out of our minds by the sheer scale of the calculations . . . To kill 278,966 civilians (this number is made up) to avoid the deaths of an unknown but probably larger number of civilians and soldiers is surely a fantastic, godlike, frightening, and horrendous act' (2006a: 262). Yet if it is merely the scale of the calculations that is troubling to Walzer, then the doctrine of double effect cannot escape his critique. Although double effect may not reach such catastrophic figures in a single instance, its iterability (its imitation and repeatability) ensures the effects of justification are illimitable. In short, if double effect can justify an attack in the first instance, it can, in principle, be repeated *ad infinitum*. Walzer's justification for negative unintended civilian deaths and injuries can potentially (and indeed has) become a precedent applied to unlimited future cases.

More pragmatically, it is unclear how the calculations necessary for due care to function can be effectively made on the battlefield. Walzer's argument states that military strategists must know how many civilians they are likely to harm and how valuable the legitimate target is to the war effort. This suggests that Walzer's ideal of due care operates under the assumption that military strategists have access to information on the battlefield that allows them to make calculations necessary to minimise risks to civilians. US combatants in Iraq, however, were often operating under the acknowledgement that they would have to make strategic decisions in the absence of full knowledge. For example, the earlier example of shelling outlined by Wright highlights the lack of certainty and clarity faced by combatants at the moment military attacks are launched. As Wright acknowledges at the end of his stint as an embedded reporter, 'no one will probably ever know how many died from the approximately 30,000 pounds of bombs First Recon ordered dropped from aircraft. I can't imagine how the man ultimately responsible for all these deaths – at least on a battalion level – sorts it all out and draws the line between what is wanton killing and civilised military conduct' (2005: 438). In a similar manner to the discussion on roadblocks in the previous chapter, combatants aiming to strike legitimate targets in the ways implied by Walzer's conception of double effect do so without knowing the full implications of their actions. In fact, it is precisely this uncertainty that Walzer relies on in order to depict civilian casualties as unintended: if combatants launched an attack with full information and civilians were killed, they could not appeal to double effect. So Walzer's understanding of double effect is, therefore, premised upon the assumption that combatants must remain uncertain about the implications of their actions at the moment they launch an attack.

In contrast, the fulfilment of the good intention is treated unproblematically. In calculations of double effect, while the level of potential civilian injuries and deaths must remain unknown, it is assumed that the good intention can be achieved and the legitimate target will be hit and destroyed in the manner anticipated. In this respect, double effect rests upon an interesting fusion of certainty and uncertainty: the certainty that the good effect will be achieved coupled with the uncertainty surrounding negative risks to civilians. It is only by assuming that the good intention can be fully achieved that calculations of due care can balance the projected positive objectives against the negative risks; that is, the uncomplicated achievement of the good intention is required for Walzer's calculation to work in a practical manner. Walzer's ideal is, in part, tied to belief in the concise and precise ability of military weaponry to hit targets. This belief is increasingly couched in the language of technology and so-called 'smart bombs'. These weapons and technologies are viewed as a direct extension of intentionality: we dial in a target and technology carries out our orders precisely.[5] Zehfuss argues that the discourses surrounding these technologies are designed to convince us that our actions are fused with our intentions, that we can directly control the effects of war (2011: 561). However, the actual application of the technology points to a far more uncertain status: smart bombs are only accurate 50 per cent of the time, have wide blast radiuses and ultimately rely on the assumption that we have correctly pinpointed the target in the first instance (Zehfuss 2011: 549). Shadid's (2006) discussion on US bombing of residential areas in Baghdad on the eve of the invasion provides an example of the limitations of precision weaponry. Residential bombings were justified on the presumption that the US Military knew that key figures in the Ba'athist regime where hiding in specific areas of Baghdad. Shadid argues that this military intelligence retrospectively proved to be based upon inaccurate information and no Ba'ath leaders where hiding in Baghdad's suburbs. In fact, Wright's admission that the Marines will never actually know how many civilians were killed in their march to Baghdad highlights the problems with achieving even retrospective certainty. The idea that soldiers can know that the good intention *will* be achieved prior to a calculation of double effect is thus largely illusionary: both the positive *intended* effect and the negative *unintended* effects are uncertain at the moment of actualisation. As with roadblocks, the operation of double effect underscores the role of undecidability implicit in every act of war.

Policing with due care

Because Walzer's scales of risk and reward are premised upon assumed rather than guaranteed outcomes, even more importance is placed on how calculations of due care are made, and who presides over the sincerity of the commitment not to avoid harming civilians. Recalling that Walzer's faith in double effect is premised upon his contention that the War Convention allows us to enforce 'hard and fast standards', due care requires a clear mechanism through which risk-as-insurance of good intention can be verified. In other words, unless there is some way to enforce the hard and fast rules of due care, or the line between sincere and insincere appeals to double effect (the precise line between moral and immoral action), then the principle of double effect can have no practical import. Surprisingly, then, Walzer purposely blurs the point at which due care has been met:

> Once again, I have to say that I cannot specify the precise point at which the requirements of 'due care' have been met . . . The line isn't clear. But it is clear enough that most campaigns are planned and carried out well below the line; and one can blame commanders who don't make minimal efforts, even if one doesn't know exactly what the maximal effort would entail.
>
> (Walzer 2006a: 319)

Walzer argues that, although the precise point at which due care is achieved is shrouded in uncertainty, we will certainly know when minimal efforts have not been made. In this we find another example of Walzer's fusion of certainty and uncertainty: the uncertainty surrounding the precise point at which due care is met is tempered by the certainty that we will know if minimal efforts to minimise civilian risks have not been made. However, the concept of 'minimal effort' is itself a limited constraint in the context of war, and Walzer acknowledges that militaries will always look for a way to juggle the figures to suit their own interests (2005: 39). More importantly, it is even harder to determine if minimal standards are enforced when we consider the fact that the minimal standard varies between cases. Walzer asserts that 'the degree of risk that is permissible is going to vary with the nature of the target, the urgency of the moment, the available technology, and so on' (2006a: 156). In this way, he introduces context as a variable in the double effect ledger: the specific context of the situation directly alters the minimal requirements of due care. Walzer's acknowledgement that minimal effort is dependent upon context is important, because it suggests a fluid understanding of due care rather than a hard and fast standard. The fluid interplay between context and risk is illustrated in US Army Corporal McIntosh's depiction of the battlefield. McIntosh argues that risk and necessity change during the course of battles and this shift in context alters the mindset and behaviour of combatants: '. . . in the heat of a firefight, the calculus sometimes change. A shot not taken in one set of circumstances might suddenly become a life-or-death necessity' (cited in Filkins 2009: 91). In this sense, due care is a fluid and mobile concept that is identifiable only at the point when combatants decide to launch an attack. This flexible conception of due care increasingly complicates the questions of where lines are drawn, how we know if a particular minimal effort to reduce civilian harm has been made, and who actually judges if it has been made.

The question of who judges, or polices, the moral rules of due care is largely ignored in Walzer's argument. As such, due care has a direct problem in regard to enforceability. In practical terms, military strategists and combatants are primarily placed in charge of the enforcement of due care because they must calculate between risks and rewards, and apportion a relative value to each variable in Walzer's equation. Combatants must determine the strategic value of the target, assess the likely risks to civilians and calculate the minimal level of risk reduction required. Nevertheless, if combatants are designated with policing duties with respect to the rules of double effect, then this poses some unavoidable problems for the possibility of judging intentions. Derrida argues that the indeterminate structure of laws opens the possibility for the police to remake, rather than simply enforce, the law (2002a: 277). This understanding of policing has important implications for Walzer's depiction of double effect. In his argument, double effect comprises a fluid and flexible set of rules that can be remade illimitably, depending on how they are interpreted by particular combatants within specific contexts. Because combatants must make context specific judgements of due care, there is no single rule or singular set of rules. Instead, a chain of individual decisions determines what constitutes a minimal effort to reduce harm to civilians.

Presented in these terms, the police are effectively policing their own borders: the combatants professing the goodness of the intention are also determining the point at which the minimal requirements of due care are met. Policing, which is necessary to protect the doctrine of double effect, now threatens it from within. More precisely, the police forces charged with enforcing the minimal standard of due care can potentially interpret the standard to fit their own strategic ends. In Walzer's depiction of double effect, the killing and injury of civilians is justified on the basis that combatants did not intend to harm them. In turn, the commitment not to harm civilians is judged in terms of due care, i.e. the efforts made by combatants to minimise risks to civilians. Yet judgements of due care are principally placed in the hands of the combatants whose intentions are in doubt. Walzer's conception of moral judgement, however, does not intend for combatants to become the adjudicators of the justness of their own actions. Rather, 'ordinary people' determine if war is conducted within its proper moral boundaries (Walzer 2006a: 15). In this sense, we need to ask how ordinary people can possibly judge appeals to the principle of double effect.

Walzer's belief that ordinary people judge the morality of war is important because it suggests that the ordinary civilians who suffer the consequences of military attacks have a role to play in judging military actions. Shadid, writing from within Baghdad during the initial bombing campaign, provides an illustrative example of Iraqi judgements of US bombing. He describes how unintended consequences potentially justified under double effect were experienced by those directly affected:

> The strike came at two P.M. on April 7, two days before the capital was conquered. A single B-1 bomber dropped four 2,000-pound bombs on a cluster of homes in the wealthy neighbourhood of Mansur, where American intelligence believed Saddam and his two sons, Uday and Qusay, were hiding . . . Residents said the bombs had sucked air from homes blocks away, as if the neighbourhood, in its entirety, gasped for breath . . . The mauled torso of twenty-year-old Lava Jamal was pulled out before they arrived. Moments later, a few feet away, others found what was left of her severed head, her brown hair tangled and matted with dried blood. Her skin had been seared off.
> (Shadid 2006: 131–133)

What is interesting in this case is how the intentions guiding attacks are altered by the unintended consequences: the intention to kill Uday and Qusay Hussein is irretrievably distorted and deformed via the unintended consequences of its actualisation. In Shadid's experiences of the bombings, the intentions guiding the attacks did not necessarily matter to those affected, as the Baghdad residents were confronted, not with the intention, but with the consequences of the attack (2006: 133). Hence, the important moral issue for ordinary Baghdadis was not if the US had intended to harm civilians, but the actual harm caused to them and their neighbours. In this sense, the Baghdad residents determined the justness of the attack through the unintended consequences rather than the professed intention.

Nevertheless, Walzer does not intimate that the victims of the unintended consequences actually matter in the determination of the justness of the act. Instead, his argument suggests that ordinary people looking in at the war from the outside are in the best position to judge. This presents a difficulty, because people outside the warzone are entirely dependent upon first-hand accounts in order to gain any understanding of how events unfold in any given war. In short, ordinary people outside the battlefield must rely on those inside to provide the information necessary for them to make moral judgements. As discussed in the introduction to this book, first-hand accounts are problematic because they can only present contextually limited narratives to various publics. In turn, the contextual dynamics of first-hand accounts have implications for the type of judgements that ordinary people can make about appeals to double effect. Embedded reporters have proved to be one of the main first-hand sources able to relay information on the planning of US military attacks in Iraq to ordinary people. Hoyt and Palatella facilitated a discussion between a number of journalists on the risks and rewards of embedded reporting (2007: 97–111). While there was some disagreement on the extent to which embedding compromises journalistic ability to provide an accurate representation of events, the majority of journalists admitted that obtaining a comprehensive account of US military actions in Iraq required some dependency upon military sources. Filkins provides an example of the extent to which reporters in Iraq relied upon military information: 'A few months later, Hajji Hussein's kebab house was destroyed in an air strike. The Americans said it was a terrorist "safehouse," from which "innocent civilians knowingly stayed away," but I always wondered about that' (2009: 220). Though Filkins's reliance on the military account may seem to diminish the idea of journalistic objectivity and credibility, there are very few alternatives for gaining information on military operations in warzones. Primarily, journalists can ask military sources who are invested in portraying their actions as justified, or they can ask the local people who have survived the military action, or locals who claim to have witnessed the event. None of these options gives ordinary people any solid ground to make the judgements about due care that Walzer's principle requires. As journalist Richard Engel contends:

> You have to rely on someone who's from there, who's bringing you the tapes, and then you have to piece together what happened from accounts from the military, accounts from eyewitnesses, accounts from hospital figures, all of whom have credibility problems. You have to piece together the best you can to come up with a mosaic of what's going on.
>
> (Cited in Hoyt and Palatella 2007: 5)

Engel's argument illustrates why judgements of due care are saturated in the context through which they are presented: where the information is coming from, who is relaying the information, why they are relaying it and in what way is it relayed. In turn, we must reflect upon the context of the audience charged with making moral judgements based on first-hand information: who they are, where they come from and their understanding of politics and ethics. In Walzer's terms, information from warzones is presented within particular moral vocabularies, and we must assume that ordinary people (who have their own maximal perception of morality) can accurately extrapolate universal moral judgements from these particularist accounts. Because the information necessary to formulate judgements about actions in war is intertwined with the contexts of its representation and interpretation, there is no way for ordinary people to definitively judge appeals to double effect: there is no universal scale that is uniformly accessible to all people at all times. Walzer requires hard and fast rules to ensure that double effect remains a universal principle that is open to the judgement of ordinary people. Yet there is no minimal way for ordinary people to judge the sincerity of combatants' intentions not to target civilians.

More problematically, if it were possible for ordinary people to effectively judge appeals to double effect, their judgements would be retrospective. This is important because retrospective judgement calls the possibility of double effect into question. In double effect, Walzer conceptualises justification as prior to the realisation of any consequences: the combatant is justified by their intention not to harm civilians prior to the action, and their good intentions must be validated by calculations of due care. If an act of double effect, then, is to be justified at the moment a military operation is launched, it can only be policed by the combatants who launch the attack. However, Walzer's requirement that ordinary people judge the moral character of war fighting means that double effect cannot be judged in the present and, therefore, the fidelity of the intention not to harm civilians must be judged retrospectively. Importantly, retrospective judgement means that due care is not a barrier to the imposition of unnecessary risks on civilians, because the acts are judged after the unintended consequences have already occurred. Remembering that judgements of due care differ in every case, the condition of retrospective judgement constitutes a supplementary deferral of justification. In fact, because judgements are dependent on the context in which the act is communicated, a definite singular judgement of justness, uniformly agreed by all 'ordinary people', is perpetually deferred. The sincerity of an appeal to double effect, both differing and deferring, constitutes an instance of *différance*. *Différance*, then, founds the difference between the act of terrorism that is always unjust and carried out in bad conscience and the genuine act of double effect which is always just and made in good conscience. In this way, the morality and good conscience promised in double effect is threatened with indistinction.

Deepening double effect

In the previous sections I have demonstrated why Walzer's conceptualisation of double effect fails to justify military actions that endanger civilians. Ultimately, Walzer's understanding of double effect reduces the principle to a retroactively judged utilitarian calculation that is impossible to implement in any practical way. Yet the nucleus of double effect, the acknowledgement that certain acts of war produce unintended and unforeseen negative impacts, has a more far-reaching import in regard to explaining ethical responsibility in war. Walzer presents us with a narrow depiction of double effect that is applicable only to military activities that are likely to kill or injure civilians as a direct consequence. In this sense, double effect is geared toward ethical responsibility defined in terms of an enclosed time-space: ethical responsibility understood in terms of the immediate impacts of military attacks. In Walzer's analysis, combatants are solely responsible for the goodness of their intention to minimise the risks their actions pose to non-combatants, and they are only required to minimise the immediate effects that their actions produce. Future consequences resulting from acts of war, therefore, are removed from combatant accountability. For example, if a military action unintentionally made a plot of land unsuitable for farming, combatants would not be responsible, in any way, for the loss of the owner's livelihood. In Walzer's model, the combatant is responsible up to the point when the attack is launched and ethical responsibility is satisfied when the combatants make 'minimal efforts' to reduce direct risks of harm to civilians. He thereby presents us with a limited conception of ethical responsibility that neglects some of the most far-reaching implications of collateral damage. The purpose of this section is to illustrate how ethics as response provides us with a better understanding of the ways in which ethical responsibility is related to acts of war.

Walzer discusses the negative consequences of double effect specifically in terms of non-combatants' right to life: combatants are required to minimise the risk that their actions will kill or harm civilians. In contrast, the destruction of infrastructure and property is either ignored or seen as a moral victory because it has not resulted in the direct loss of life: 'One can destroy a great deal of property in answer to the destruction of human life' (Walzer 2006a: 218–219). The non-moral depiction of infrastructural damage exemplifies Walzer's focus on the immediate implications of military action. In short, he views damage to the infrastructure as a secondary concern that does not pose a major barrier to ethical satisfaction in war.[6] Nonetheless, infrastructural collateral damage is intimately related to long-term impacts on human life. For example, in the context of Iraq, the impact of bombing during the Gulf War of 1990–1991 drastically altered civilian life for over a decade, and continued to shape civilian lives under US occupation and beyond. Journalist Rajiv Chandrasekaran explains that targeted US bombing during the Gulf War damaged about 75 per cent of the country's power generating capacity, crippling the civilian power supply (2008: 167). Electricity is a crucial good in Iraqi society needed for water treatment, powering hospitals, along with domestic cooking, heating and lighting. Gregory spells out the devastating effects of the destruction of Iraq's electrical infrastructure: 'Without power, water-treatment and sewage facilities shut down, and thousands of people (particularly children) died from diarrhea, dysentery and dehydration, gastroenteritis, cholera, and typhoid' (2004: 168). As such, infrastructural damage is directly related to future risks to civilian lives.

The targeting of infrastructure is often a key component of military strategy because it potentially cripples the mobilisation capacity of enemy troops. Recognising this, Walzer makes a distinction between goods that have a specifically military purpose and goods that are pivotal to the wellbeing of the civilian population. Walzer argues that the targeting of civilians who work in sectors such as power generation is unjustified because they make goods that are needed by both military personnel and civilians (2006a: 146). This is an acknowledgement that those who make the goods needed by the civilian population are immune from attack. Yet Walzer does not propose any moral barriers to the targeting of infrastructures necessary for the production of civilian goods. In other words, militaries must refrain from attacking civilians producing the goods that a civilian population needs to survive, but they have no parallel duty to avoid targeting civilian infrastructure. This idea of responsibility was echoed by US military actions during the Gulf War. Shadid explains that US targeting of Iraqi infrastructure was intentional and justified upon a utilitarian calculation: 'The choice of these targets was justifiable; their losses would incapacitate the Iraqi army, recognised as an aggressor by the United Nations.' (2006: 44). In turn, Gregory contends that the US Military targeted infrastructure during the Gulf War despite the obvious dangers to the civilian population: 'The U.S. intelligence agency had estimated that "full degradation of the water treatment system" in Iraq would take at least six months, and its destruction would cause serious public health problems' (2004: 168). Despite the explicit understanding that the destruction of infrastructure poses major risks to civilian life, as illustrated in the example of Iraqi electricity, Walzer never directly discusses the destruction of property in terms of double effect; thus, the protection of infrastructure is not deemed to be a moral imperative. Acknowledging that targeting infrastructure, intentionally or unintentionally, puts civilians at risk begins to explain why the ethical implications of military actions cannot be conceived in terms of a bounded timeframe. The effects of infrastructural bombing during the Gulf War, as discussed in the second chapter, extended past the 2003 intervention into the occupation and beyond, fundamentally limiting the capacity of the US to provide security and services to the local population. In this sense, the destruction of infrastructure is never simply the loss of property or services as Walzer's theory implies: it can pose a clear long-term threat to the security and wellbeing of various populations.

Siege warfare: an illustrative example

The targeting of infrastructure during the Gulf War is just one example of the importance of looking at the temporally unconstrained effects of warfare.[7] The underlying argument is that it is not possible to cut responsibility off at a specific point on the grounds of good intention, or adherence to moral rules. There are two primary reasons why we cannot de-limit ethical responsibility in this way: first, military actions are iterable and, therefore, always in the process of becoming other than intended; and second, the consequences of military actions engender future contexts that, to a certain extent, dictate the ways in which future ethical relationships can take place. In this sense, it is not possible to derive ethical satisfaction through compliance with rules like double effect because we do not ever fully know how the intended and unintended effects impact other people. The siege of Fallujah provides an illustrative example of the implications of cutting responsibility off at a singular point as Walzer's conception of double effect suggests.

Walzer describes siege as an instance in which combatants attempt to shelter themselves in a city or town among civilians in the belief that their enemies will relent due to fear of mass civilian casualties (2006a: 160–161). Siege poses problems for the principle of non-combatant immunity because it places the forces laying siege in a position where they can only attack their enemy by intentionally endangering a civilian population: the besieged combatants have taken refuge among civilians and their opponents would therefore have to undertake great risks if they attempted to ensure the safety of civilians. Again, the dual imperatives of winning the war and protecting civilians are placed in direct conflict. Walzer attempts to resolve the tension between the imperatives by stating that combatants laying siege to the city or town are absolved of ethical responsibility if they offer civilians safe passage: 'The offer of free exit clears him of responsibility for civilian deaths' (2006a: 169). Walzer's argument has two components: first, if civilians are forced to remain in the city/town, then those who force them to remain are responsible for the danger they face; and second, if civilians freely choose to remain in the city/town they forfeit their civilian rights. Walzer asserts that civilians who choose to remain, or are forced to remain, have been effectively conscripted into the besieged garrison and have therefore 'yielded their civilian rights' (2006a: 168–169).[8] Once more, Walzer wants to impress forfeiture in terms of individual choice: if civilians refuse the offer of free exit, or are denied exit by occupying troops, they forfeit their rights.[9] While Walzer implies that the potential 'yielding' of civilian rights in times of siege is regrettable, his central argument is that the combatants laying siege are not responsible for the risks imposed on civilians. If combatants offer free passage to those inside the besieged city or town, they are absolved of their ethical responsibility to refrain from killing non-combatants; responsibility is cut off at the point at which the moral rule is followed.

The second siege of Fallujah in November 2004 helps illustrate Walzer's logic. Fallujah had an estimated population of 350,000 to 500,000 people and is considered one of the most important holy cities in Iraq. In April 2004 conflict sparked when Iraqi civilians were killed by US troops during a protest against the closing of a school. In response, Iraqi resistance fighters killed four US contractors working for the Blackwater security company. The resistance fighters then dragged the mutilated bodies of the contractors through the streets and suspended them from a bridge (Chan 2004). The US took swift action by laying siege to the city. The April siege symbolised a collective punishment of Fallujah residents for the crime of harbouring resistance fighters. In the words of US Brigadier General Mark Kimmit, '[c]ollective punishment is imposed on the people of Fallujah by those terrorists and cowards that hunker down inside mosques, hospitals and schools' (cited in Holmes *et al.* 2007: 123). The initial siege ended in a stalemate with the policing of the city handed over to the local Fallujah Brigade comprised entirely of Iraqis. Throughout the summer and autumn of 2004 US intelligence identified Fallujah as a resistance stronghold. Consequentially, US Marines decided to launch a second siege with the aim of trapping a large volume of resistance fighters within an enclosed space. According to First Infantry Lieutenant Colonel Pete Newell: 'We don't want them to leave Fallujah. We want to kill them here' (cited in Lasseter and Allam 2004). Nonetheless, in line with the humanitarian motivations proclaimed in the Bush Administration's justification of the war and occupation, the US wanted to demonstrate that they were taking precautions to minimise civilian casualties. The US strategy in Fallujah resembled an abridged version of Walzer's model of siege warfare. The Marines offered free passage to all women, children and elderly men (Jamail 2008: 135, 234). Families were officially given seventy-two hours to leave the city or be designated as legitimate targets (Holmes *et al.* 2007: 76). Similarly to Walzer's argument, the US Military viewed the offer of free passage as an absolution of the principle of non-combatant immunity. In the words of US Marine Sergeant Medows, '[w]e had dropped flyers a couple of days prior saying to people to get out of the area if they didn't want to fight, so basically anyone who was there was a combatant' (cited in Holmes *et al.* 2007: 119–120). Yet there were complications with the implementation of the strategy. For example, the majority of women in Fallujah were unable to drive and were therefore unable to leave the city unless their husbands drove them. With men barred from leaving the city, numerous women were also forced to remain (Holmes *et al.* 2007: 58). More importantly, many civilians did not trust the offer of safe passage because civilians waving white flags had been shot by US snipers during the April siege (Jamail 2008: 250). The Iraqi Red Crescent claimed that they knew of at least 157 families still trapped in the city at the time the siege was launched (Holmes *et al.* 2007: 68–69). The US was reluctant to discuss numbers of civilians that remained in the city during the siege. After a number of months, however, US officials finally acknowledged that between 30,000 and 50,000 civilians had remained in the city (Jamail 2008: 234).

Walzer legitimises siege warfare in terms of a moral imperative: the offer of free passage absolves troops laying siege to the city/town of all ethical responsibility. Again, Walzer's argument presupposes a homogeneous model of communication in which troops inform civilians of their right to free passage and this message is unambiguously understood, i.e. communication operates in a linear uninterrupted manner: the intended message is fully understood by its intended audience. The case of Fallujah, however, highlights why the specific context of a siege impacts upon how the offer of free passage is interpreted by the people it is directed toward. The Derridean depiction of communication as always in the process of becoming other than its intended meaning helps explain why Iraqi civilians remained in Fallujah. The US offered conditional free passage to Fallujah residents, yet the interpretation of this offer transformed the siege into an instance in which the targeting of non-combatants was morally justified on the basis of misinterpretation rather than the free choice of Fallujah residents. Fallujahians did not necessarily understand or believe the offer of free passage, and many civilians could not actually leave despite wanting to. The net result was that a large number of civilians remained in Fallujah during the siege. As this example illustrates, Walzer's linear depiction of communication does not sufficiently account for the ethical questions raised by siege warfare. His model presupposes that the decision to remain in the besieged city/town is either inflicted by the occupying force or is the free choice of civilians. Fallujah highlights why the ideal of free choice is inadequate because it ignores the iterable and transformative dimension of communication. Siege warfare, therefore, provides another example of the implications of Walzer's desire to cut responsibility off at the point where a moral rule is followed: responsibility is fulfilled in the singular offer of free passage.

Although the communication of free passage is problematic in itself, this is not the biggest problem in Walzer's argument. By cutting responsibility at the point where free passage is offered, Walzer is also suggesting that combatants are absolved of all future responsibilities to the civilians who leave the besieged city. He acknowledges that forcing civilians to become refugees is lamentable, but nevertheless maintains that this does not render siege warfare morally unacceptable (2006a: 169–170). The consequences of free passage, however, overspill the initial context and impact upon future events. The Fallujah example demonstrates why creating a vast population of refugees can potentially create new ethical problems. In response to the US offer of free passage, 203,000 of Fallujah's 300,000 residents fled the city (Herring and Rangwala 2006: 181). The first impact of this was that many of the Fallujahian refugees were unable to return to the city:

> According to the official estimate 'almost 36,000 houses have been demolished, 9,000 shops, sixty-five mosques, sixty schools, the very valuable heritage library and most government offices. The American forces destroyed one of the two bridges in the city, both train stations, the two electricity stations, and the three water treatment plants. It also blew up the whole sanitation system and communication network'.
>
> (Holmes *et al.* 2007: 21–22)

The official estimation of the damage inflicted by the siege depicts a city that, for all practical purposes, had been rendered uninhabitable. As such, siege does not mark an enclosed epoch of the war, an event that punctuates civilians' return to normal life. The consequences of siege meant that many Fallujahians could not resume their lives in the city, and by May 2006 one-third of residents were still unable to return (Holmes *et al.* 2007: 24). In this sense, the idea that responsibility ends with the offer of free passage is simplistic, because it assumes that civilians can return to safety once the siege has ended. In addition, the migratory dynamics of those fleeing besieged cities/towns can have major socio-political impacts. In the case of Iraq, Fallujah's refugees significantly contributed to the escalation of ethno-sectarian tensions recounted in the second chapter. The vast majority of Fallujah's population were Sunni. In turn, a sizeable proportion of Sunnis displaced from Fallujah were invited to live with relatives in other Sunni dominated cities, towns and villages. Difficulties arose because there was nowhere to house those displaced, and this led to concentrated campaigns to evict Shi'a minorities from Sunni majority towns. Journalist Nir Rosen describes the impacts in the Anbar province: 'That's when ethnic cleansing really got started. The first stories you heard of Shias being pushed from their homes, of getting letters, of their homes being bombed' (cited in Hoyt and Palatella 2007: 95). In this way, the displacement of Fallujah residents contributed to the escalation of sectarian violence that marked Iraqi society in the following years. Indeed, the widespread Sunni boycotting of the Iraqi election and rejection of the political process was partially a response to the destruction of Fallujah (Allawi 2007: 340).[10]

The offer of free passage, therefore, does not resolve the ethical problems associated with siege warfare. Free passage and the intentional creation of mass refugees from Fallujah produced its own chain of unforeseen consequences. The attempt to limit responsibility to compliance with moral rules is a fiction, because the consequences of action can always overspill its original context. The principle point is that acts of war must be viewed within a broader remit that extends beyond immediate impacts. In other words, we must account for the unintended and unforeseen consequences that acts of war produce: the effects of actions overflow their immediate foreseeable – intended and unintended – impacts. Because we must knowingly risk unforeseeable consequences when we act, we must also accept responsibility for these consequences. This reconceptualisation of double effect creates a necessity to sustain de-limited responsibility. We are no longer responsible up to the specific point where we have followed the moral rule, because the consequences of our actions are projected into the future.

Conclusion: ethics as double effect

Thus far this chapter has discussed double effect in terms of direct military actions. This is unsurprising given the narrow scope of double effect as defined by Walzer. Because double effect is concerned with foreseeable and calculable negative risks, it must necessarily focus on military activities that are likely to produce negative impacts in all contexts. A bomb dropped on a heavily populated area, for instance, is likely to kill or injure people. However, by deepening double effect beyond the direct and immediate impacts of an act, we have the opportunity to conceive of a new typology. In this typology, unforeseeable and uncertain consequences become morally relevant, and this allows us to consider other actions associated with war and peace-building as acts that produce double effects. Since all wartime actions potentially produce negative unforeseeable consequences, this means that understanding the ethical implications and responsibilities attached to actions in war requires us to be attentive to the way actions shape and reshape socio-political and economic contexts. Rather than reducing ethical responsibility in war to compliance with a rule set or principle, as in Walzer's depiction of double effect, we must account for the mutations and transformations of the consequences, intended and unintended, that arise within the specific contexts created by acts of war; that is, we must account for how acts of war effect (and affect) people in the future. This typology illustrates how the idea of just war can be understood as a reconceptualisation of double effect. Just war thus necessarily entails the risk of negative unintended and unforeseeable effects in the pursuit of its purportedly justified aims.

In this respect, the Iraq War and occupation have played out within the eye of a myriad of unintended consequences stemming from both direct military actions and activities more closely associated with peace-building exercises. Unintended consequences have resulted from military strategy. The US bombing of northern Iraq in 1998, for example, resulted in mass arrests of those hostile to the Ba'ath regime, which ultimately led to a vacuum of US intelligence within Iraq. Ricks explains that the resultant information gap was exploited by exiles such as Ahmed Chalabi who fervently testified that Saddam Hussein possessed WMDs (2007: 19, 57). In turn, the assumed existence of WMDs ensured that the US Military refused to detonate bunkers containing weapons caches due to fears that they would trigger WMDs. These weapons caches subsequently became the main source of armament for the resistance (Ricks 2007: 145–146). Such singular decisions thus erupt into a chain of uncontrollable and unintended consequences. Unforeseen consequences have also stemmed from US peace-building attempts to win 'hearts and minds'. For instance, combatants attempting to build relationships with local Iraqis put civilians in danger because they were viewed as occupation sympathisers (Finkel 2011: 39).[11] Reconstruction projects also contained an implicit element of double effect because they were targeted by resistance fighters as a means to demonstrate the inability of the US to improve Iraqi infrastructure. In Filkins's words, '[a]nything the Americans tried there turned to dust. The Americans repaired a brick factory and insurgents blew it up. The Americans painted a school and the insurgents shot the teachers' (2009: 82–83). Chandrasekaran claims that journalists were not even allowed to report on reconstruction efforts due to fears that projects would be destroyed or that the locals involved would be targeted (Hoyt and Palatella 2007: 136).

The examples listed above illustrate why any wartime action is an instance of double effect: we aim at a positive result but risk negative unintended and unforeseeable outcomes. The US, for instance, argued that their intervention in Iraq was justified because it would save Iraqis from the violent tyranny of Saddam Hussein. However, the toppling of the Ba'ath regime risked the creation of new violent threats and new tyrannies. Double effect resembles an alternative iteration of the sacrificial risk implicit in every ethical action. The good intention guiding the action necessarily and simultaneously entails the risk of unintended and unforeseeable negative impacts; responding to one hardship risks the unintended production of another hardship. In this sense, double effect mirrors the coupling of responsibility and irresponsibility implicit to the idea of ethics as response. The main purpose of this chapter has been to demonstrate why the principle of double effect, as presented by Walzer, is incapable of justifying collateral damage. Yet I have also sought to explain why the unintended and unforeseeable consequences implicated in the doctrine can help us understand the transformative character of war. Walzer wants to impress a conception of double effect where the intention to do the right thing absolves combatants of ethical responsibility with regard to the consequences of their actions: actions are justified in relation to intended, rather than actualised, ends. This chapter has demonstrated that iterability forecloses the possibility of intentionality working in the way that Walzer requires. Any potential action must risk consequences other than those intended, and because this risk is part of the structural possibility of all actions, the risk of unintended consequences cannot be excluded from the analysis. Ultimately, Walzer's conception of double effect is supplanted by the utilitarian calculation of due care in which the goodness of the intention is validated through the minimisation of risks toward civilians. Walzer proposes a system in which the value of the target is balanced against the projected risks imposed on civilians. Because his system requires combatants to possess direct knowledge of the risks imposed on civilians, these risks must necessarily be intended. In the absence of knowledge of potential risks, the calculations necessary for due care to function cannot be undertaken. Walzer's conception of double effect therefore leaves us with a rule utilitarian calculation conducted by the combatants who are plotting actions. In turn, the fidelity of the intention to minimise risks to civilians can only be judged retrospectively. According to Walzer's argument, judgements of due care differ between cases and judgement is deferred; in other words, the doctrine of double effect is underscored by *différance*.

Nevertheless, the unintended and unforeseeable consequences evident in acts described as instances of double effect tell us something important about ethical responsibility. Double effect constitutes the necessary structure of all action: the actor aims at the intended effect but can only do so by risking unintended consequences. In this respect, double effect parallels the fusion of responsibility and irresponsibility emphasised in ethics as response. We can only enact our responsibilities toward others by risking unforeseen, potentially negative, effects. Looking at the temporally expanded consequences that emanate from acts of war helps us to understand why ethical responsibility cannot be satisfied via moral rules or appeals to good intention. Moral rules and good intentions can never fully control the outcomes they produce. Walzer does not deny this aspect of war, maintaining that war regularly creates 'unpredictable, unexpected, unintended, and unavoidable horrors' (2005: 155). Because war risks changing the socio-political constructions within a community, those implicated in acts of war need to maintain their limited responsibility in relation to the new social contexts they help produce. The history of Iraq outlined in Chapter 2 spells out the dynamics of this interplay between good intentions and negative consequences. In attempting to solve the problem of war via an appeal to moral rules and norms we ignore the actual consequences that emanate from our actions. In this respect, the reliance on moral rules ignores both the ethical and political dimensions of violence. It ignores how war and violence transforms societies and shapes the contexts in which future ethical relationships take place.

Notes

1 This allusion to the supplement should be read as analogous to the Derridean concept outlined in the previous chapters. In fact, it will be further explained how the concept of due care threatens to usurp the centrality of intention in regard to double effect.

2 On this point I am unsure as to whether the irony is intentional on Walzer's part. But it is interesting, nonetheless, that Walzer implies that officers can appeal to the principle

of double effect, which we must remember holds intention as its governing centre, in a thoroughly unintentional manner.

3 The COIN manual provides a slight variation on Walzer's formulation: the expected damage to civilians must be balanced against the harm the target would cause if allowed to escape (Petraeus 2007: 247–248).

4 See *Just and Unjust Wars*, Chapter 1.

5 For a full discussion on precision bombing see Maja Zehfuss (2011), 'Targeting: Precision and the Production of Ethics', *European Journal of International Relations*, 17(3), pp.543–556.

6 It should be noted that Walzer also ignores the way people relate to property, and its cultural, social and historical significance. For example, Sheikh Hamed, an Iraqi farmer, recounts the destruction of his orchard in terms of historical and familial mourning: 'These are our grandfathers' orchards . . . This is our history. When they fell a tree, it is like they are killing a member of our family' (Jamail 2008: 261).

7 More than 375,000 Iraqi university students, for instance, were unable to continue with their studies due to bombing damage during the 2003 invasion (Chandrasekaran 2008: 3).

8 In many respects the offer of free passage more clearly resembles a direct threat: 'Leave the city or you will be killed.' It is important to remember, in this context, that Walzer maintains that no person can be threatened unless they have forfeited their rights through some identifiable act (2006a: 135). Thus, in siege warfare, Walzer permits combatants to directly threaten civilians on the basis that their adversaries have chosen to fight among them.

9 It must be noted that this also signifies another instance in which the right to life is potentially lost due to an infringement of the right to liberty: civilians can be targeted because they have been forced to remain in the city/town.

10 Subsequently, the US Military acknowledged that many resistance fighters had already vacated the city in anticipation of the siege. Ultimately, four days after the siege began, resistance fighters captured the much larger city of Mosul, with 3,200 of the city's 4,000 police officers deserting their posts (Ricks 2007: 304). As such, the strategic imperative also proved largely ineffective.

11 This complicates the COIN directives that troops immerse themselves in local life and live among the population (Petraeus 2007: 40, 239). How can troops embed themselves in a population with the express aim of protecting that population when doing so increases the likelihood of civilian deaths?

Conclusion

Ethics as response

What the hell is water?

Walzer's central argument is that his theory of war offers us tools that clear up the confusion and messiness that we are faced with when we start to think about what it means to act ethically in war. In his terms, the pressure and immediacy of the decisions we face during times of war rob us of our capacity for patient reflection. Walzer's rules, in this respect, provide clear and accessible guidance to help us act in a morally justified manner under the duress of war: when we are faced with a troubling decision, that will most likely place lives in danger, we can consult Walzer's rulebook and it will help us clarify what the correct response should be. This is an understandable aim, and is certainly not an entirely negative aspiration. Most people want to gain the satisfaction derived from doing the right thing and acting in the right way, and conventional rules help us come to terms with difficult decisions by providing a safety net. Yet this desire to know what the correct response is, and thereby satisfy our desire to know that we have done the right thing, weakens our active political engagement with the world around us. The authoritative security of rules and guidelines allows us to switch off our thinking; it enables us to stop thinking about how our actions may affect other people and get on with the mechanical task of implementing the established rules. Adherence to rules depoliticises the problems we are faced with by assuring us that the problems were never really problems, because they have already been resolved by someone else at an earlier time.

David Foster Wallace, in a 2005 keynote address to graduates from Kenyon College, provides an interesting discussion on why resting upon conventional assumptions disengages us politically. Wallace recounts a didactic story of an encounter between two young fish and an older fish. As the two young fish are swimming around, the older fish enquires, 'How's the water?' The two younger fish swim on for a bit before asking each other, 'What the hell is water?' The point behind Wallace's parable is that the things we tend to engage with the least critical reflection often embody some of the most pressing aspects of our realities. For example, the notion that it is morally acceptable to kill certain people in certain circumstances underpins all justifications of war, and we often simply accept this fact because it seems so intimately related to the practical reality of war.

How could it be otherwise, when war suggests that at least some of those doing the fighting risk death? In contrast, Wallace suggests that questioning the seemingly uncomplicated aspects of our reality can lead us back toward a more active political engagement with the world around us. Primarily, he is arguing that being politically active means resisting the urge to unconsciously accept that some things must be understood in a specific way – that is, if we critically examine what seems so obvious and so definitive, we start to develop a deeper relationship with our capacity for critical thinking and political engagement. For instance, in regard to the killing of combatants, we might want to think about the specific consequences that could arise if an individual is killed during war: how it could affect their loved ones, how it could set a precedent that enables someone to justify further killing in the future, how it could motivate retaliatory violence, and so on. Questioning the seemingly obvious aspects of reality, in this sense, puts us back in the midst of ethical relationships: we are no longer thinking about the rules; we are, instead, reflecting upon how our actions might affect other people. Wallace argues that positive aspects of active engagement unfold in a plethora of, often minor and seemingly inconsequential, ethical relations: 'The really important kind of freedom involves attention and awareness and discipline, and being able truly to care about other people and to sacrifice for them over and over in myriad of petty, unsexy ways every day.' What Wallace means is that even though our actions are often minor and seemingly insignificant, this does not make them any less ethically or politically important. Ethical responsibility, in this respect, means thinking critically about the seemingly inconsequential actions that comprise our daily engagements with other people.

The key point in Wallace's discussion is that we need to remain critically engaged with the world around us and reflect upon how accepted understandings of the world facilitate particular responses to ethical and political questions. Conventional responses to the questions raised by war are not inherently wrong, or even drastically misguided. They are often thoughtful and articulate, and this is part of the reason why they can seem so attractive. Robust models of conventional ethics are carefully constructed by intelligent people with admirable aims. Conventional models, nevertheless, nourish the idea that we can switch off from our ethical engagement with other people. They suggest that we can resolve ethical problems, in some ways and some respects, if we stop thinking critically and commence the bureaucratic administration of justice. This is part of the reason why Walzer started writing about war in the first place: he was unhappy that people had resolved to think about war in a particular way, and attempted to provide a platform through which ordinary people could start thinking critically about war again. The purpose of this project has been to demonstrate that Walzer's attempt to give discussions on war a fresh start remains ensnared by the ideal of resolution: if we find the right rules, we can solve the problem of war. Conventional models of ethics, in this respect, purport to offer salvation: following the rules is tantamount to acting in the morally right way. The aim of ethics as response is to highlight the need for a critical reconceptualisation of the accepted ways in which we begin to think about questions of ethics, justice and war. This conception of responsibility calls for us to face the challenge of the unflinching ethical hold other people have over us at every moment. Ethics as response asks us to see the ethical dimensions of war in terms of a daily political engagement. It demands that we resist our desire to solve war's ethical questions once and for all, and accept the challenge of responding to a call for justice without the possibility of satisfaction.

The overarching aim of this book has been to illustrate why it is impossible to find clear solutions to the ethical questions posed by war. Walzer does not necessarily disagree with this contention, describing war as an unending debate; in this respect, he does not view his rulebook as the last word on war, but as a starting point for a sustained political engagement. Walzer's argument, however, starts from the premise that we must accept a set of universal truths before we can begin to debate the morality of war: we must acknowledge that human beings all share an underlying moral reality that is unproblematically represented by human rights discourse; we must accept that that it is morally permissible to kill certain individuals in war without sacrificing their rights; and perhaps most importantly, we must recognise that self-determining states, separated by established borders, constitute the authentic mode of human existence. Yet Walzer's universal truths, in many respects, represent some of the most important ethical and political questions that arise in discourses surrounding violence and justice. His foundation is, in this sense, loaded with ethical and political assumptions framed as unproblematic ontological facts. Most problematically, Walzer's foundation presupposes that the question of what justice means has already been resolved. He points toward an image of justice that is inseparable from his communitarian understandings of politics and society: a just war is a war that protects or reconstructs a self-determining state. He thus presupposes that we have resolved the biggest question before we even begin to discuss the relationship between violence, ethics and justice. Walzer's certainty that we already know what justice means, therefore, helps him to resolve much of the ethical complexity and messiness bounded up in questions of war.

The previous chapter concluded by arguing that the possibility of any ethical action is underpinned by the notion of double effect; the actor aims toward a positive intention, but risks a myriad of unintended and unforeseeable negative impacts in the pursuit of this end. In turn, Walzer's understanding of a justifiable war is also underpinned by the notion of double effect: those responding to aggression or acts that shock the moral conscience aim toward a desirable effect – a just resolution – but can only achieve this result by risking the horrific consequences of war. In some respects, Walzer acknowledges this depiction of war. War is hell precisely because people suffer and it forces them to risk their lives. Nevertheless, Walzer tempers the ethical uncertainty implicated in the resort to war through an appeal to moral certainty: we can risk this suffering because we are fighting for justice. His belief that we have already resolved the question of justice, therefore, underpins every facet of his moral argument. In contrast to Walzer's taming of uncertainty via an appeal to justice, ethics as response maintains a conception of responsibility in which the possibility of justice is definitively precluded. The fulfilment of justice is impossible because we never fully know how our decisions and actions will impact upon other people or how they will shape the future contexts that they face. In this sense, it is the fact that we are not fully in control of the consequences of our actions that denies the satisfaction of justice: we cannot justify our actions on the basis of our intention to make things better, because we cannot control how other people respond to the new contexts that our actions help to create. This, as I have explained in the previous chapter, signifies the becoming other of intention through action. Ethics as response, in other words, points toward an understanding of war in which justice cannot be guaranteed in the decision to go to war. Instead, war signifies an ethical engagement in which the realisation of justice is perpetually deferred through the unintended and unforeseeable responses that the decision to engage in violence in the name of justice initiates. To be responsible, therefore, means to remain involved in all the politics and messiness that follows our engagements with other people.

Shattering Sisyphus

In the conclusion to his discussion on ethics and international politics, Dan Bulley presents the mythological narrative of Sisyphus's punishment as an allegory for Derridean ethics (2009: 112–113). Bulley contends that the image of Sisyphus ceaselessly pushing a boulder toward the crest of a steep hill only for gravity to drag it back to the bottom before it traverses the peak symbolises the impossibility of justice: we orient our action toward the impossible ideal in the knowledge that our efforts will always fall short. Although the image of Sisyphus captures the desire for an active and sustained engagement with action, the metaphor is still captivated by the ideal of control. Bulley's metaphor is underpinned by a routine familiarity: Sisyphus is always pushing the same boulder up the same hill, aiming toward the same impossible goal. In this respect, Bulley presents a normalised notion of ethics: we know what to expect from ethical action before we even begin, and our ethical failures lead us back to a familiar starting point. Perhaps most importantly, the metaphor excludes the notion of the ethical relationship itself; Sisyphus is the only actor involved, he alone pushes the boulder, it is his burden and justice is his deferred reward.

What I would like to suggest is that the impossibility of justice is not premised upon our inability to push the boulder over the peak; it is not the consequence of an impossible horizon. Instead, justice is an experience of the impossible precisely because we always push the boulder over the peak, and we cannot avoid doing this. When we act in the name of ethics, we push our boulders toward justice and toward our desired outcomes. However, at the very moment we believe that we have completed our task, we lose control and the boulder cascades over the edge – our actions always go further than we would like them to. We want to conceive ethical action in a way that allows us to achieve satisfaction and cessation. We want to perch our boulders on top of the cliff and bask in the satisfaction of accomplishing our task. We want to know, definitively, that our actions have accomplished a clear resolution, that they have reached a resting point.

Yet this ideal of stasis represents what I believe is a better figure of impossibility. It is never possible to reach the satisfaction of stasis because it is never possible, in all rigorous certainty, to know when our actions have ceased to produce impacts. In this respect, ethical satisfaction is beyond the horizons of our thought – not because our actions always fail, but because our actions set in motion events that are never fully under our control and always risk becoming other than we intended. As such, the impossibility of justice is not that success is beyond our reach; rather, we cannot achieve justice because we cannot fix the outcomes of our actions. In pushing our boulders toward the peak, we are pushing them toward other people and toward an unforeseeable future that lies beyond our control.

In this way, ethics as response is an acknowledgement that justice is deferred because we don't know how our actions will impact upon other people. When we act, other people are affected: sometimes physically, sometimes emotionally, sometimes ideologically, and so on. Our actions may result in dramatic changes to social and political contexts. At other times, they may become a rallying point for political activism. Most often, our actions will have what we describe as minor or negligible impacts. But what matters is not the scale or immediacy of the impact. Instead, it is the fact that we cannot guarantee how other people respond to our actions; in other words, the action is not an end in itself. The action opens itself toward other people who are free to react to our actions in a multiplicity of different ways. We push the boulder over the cliff; it falls to the earth and shatters into the possibility of infinite interpretations, reactions and responses. This makes the question of ethical responsibility messy and complicated. It is no longer the question of how a singular subject can act in a morally justifiable way toward others; it is a question of the relationship between the subject's singular action and the myriad of responses and consequences that follow this action. Yet, as Nancy reminds us, the ethical relationship never exists abstracted from this chain of response: 'there is only an infinity of shatters' (1991: 101). In this respect, to act ethically means to be faced with a multiplicity of potential responses, all of which risk an infinite number of unforeseeable consequences. Our limited capacity for action means that we have to choose between multiple responsibilities and multiple actions that can affect other people in innumerable uncontrollable and unforeseeable ways.

This is the understanding of ethics proposed by the idea of ethics as response. It is a call for us to remain connected to the consequences of our actions, and the way in which other people respond to the contexts we help create. This conception of ethical responsibility resists the notion that the ethical act completes the circle and becomes an end in itself. It rejects the idea that acting in the right way is the rightful end of our responsibilities to other people. Ethics as response, instead, asks us to see ethical responsibility in terms of a sustained political engagement, a movement between action and response: a chain of responses without the possibility of stasis. It is an idea of ethics that takes the form of a reflective critical engagement with the daily decisions we all face, and the consequences that follow from the decisions we make. Taking account of consequences – of the new contexts our actions help create – is the nucleus of the Derridean understanding of justice as an experience of the impossible. There is no way that any one person can respond to all the ripples their actions create and, in this sense, a sustained ethical engagement requires subsequent decisions and responses, further chains of action, consequence and response. In regard to Walzer's ideal of empowering ordinary people, ethics as response remains a call for people to start thinking about the ethical questions posed by war. Unlike Walzer's model, it does not suggest that we need to accept some fundamental assumptions, or that we need a common language to start thinking about war, or that we can stop thinking critically about these questions. Ethics as response does not offer the imminent horizon of justice or resolution. In contrast, it asks us to remain troubled by war and to keep asking questions about why we are sometimes comfortable with the prospect of men and women being killed on the battlefield in the name of some greater good.

Responding to Iraqis

The US intervention in Iraq was predicated, in part, upon the idea that violence was justified because it promised the creation of a free and safe Iraq. The 2003 invasion, therefore, highlighted the complicity or co-option of human rights discourses within the mechanics of contemporary Western violence: violent intervention in Iraq was justified as a defence of Iraqi rights. In another respect, the invasion articulated the deep tensions contained within human rights discourse. Iraqi rights needed to be protected from Ba'athist oppression, but protecting Iraqi rights necessitated risks to Iraqi lives and the lives of coalition combatants. Thus, the contemporary Western ideal of justifying violence in the defence of rights is bounded within the logics of sacrifice: rights can only be protected if we are willing to risk certain lives. The aftermath of the 2003 invasion and occupation underscores the risks that accompany the desire to protect or liberate people through violence. Not only were lives lost during the initial period of fighting, but the invasion culminated in a nation beset by deep social, political and economic problems, and dramatic ethno-sectarian divisions punctuated by violence. The current socio-political contexts in Iraq, however, are not simply the result of neoconservative ideology or poor military strategy. Instead, the analysis of post-war Iraq presented in this book highlights that the way people (both academics and non-academics) commonly think, talk and write about war and violence is implicated within the fabric of Iraq's violent post-war reconstruction. Conventional responses to the questions of war, ethics, justice, politics and community have created discursive contexts in which particular responses have become more likely. For example, one of Walzer's core arguments is that constructing social and political unities can advance people's security. In turn, this idea was a central component of the US plan for post-war Iraq in which political representation was premised upon homogenous ethno-sectarian groupings. Ultimately, this ideal of security through unity has fuelled conflict within Iraq to the point that all three main ethnic groupings in the country (Shi'a, Sunnis and Kurds) are, in various ways, attempting to cement a clear sphere of political rule underpinned by ethno-sectarian unity. Dominant discursive frameworks, in this way, provide a canvas upon which certain responses and certain actions appear more appealing: our most prevailing images of reality tend to reproduce themselves.

In more practical terms, the attempt to liberate Iraqis from tyranny has, in a number of ways, exacerbated the daily risks of violence in Iraq. The US intervention, in this respect, has proven a resounding failure – and this failure may have the effect of solidifying the ideal of self-determination. Because foreign engagement negatively impacted upon Iraq and its population, we should just let *them* get on with the task of sorting out their own country and its politics. As such, the failure of the US occupation in some respects makes Walzer's arguments even more appealing and convincing. This book, nevertheless, maintains that the persistent problems of violence in Iraq should not frighten us away from further political engagement with Iraqis and their emerging society. It rejects the claim that our previous failures should push us toward disengagement, and that we should just let Iraqis get on with rebuilding their own country. Ethics as response thereby rejects the assumption that self-determination will necessarily resolve Iraq's problems. Instead, it calls for a reinvigorated political engagement with Iraq that incorporates the different modes and conceptions of ethical responsibility outlined throughout this book. The question of how non-Iraqis can respond to the ethical problems that have arisen in Iraq hits upon what I believe is an important failing in the way people conceive ethical and political engagement. When non-Iraqis are asked how they would respond to Iraq they generally imagine sweeping and immediate impacts; the idea of dramatically changing Iraq for the better with a few major actions. However, this image of ethical action feeds back into the ideal of disengagement. It suggests that if the sweeping changes can be achieved, people can once again forget about Iraq. By fulfilling our responsibilities in the most direct way, we can solve the problems once and for all. In contrast, I would like to suggest a model of ethical responsibility that embraces what Wallace describes as the 'unsexy' political engagements that we encounter on a daily basis. Ethics as response suggests that we should not respond to Iraq – the ideal of a unified nation – but instead respond to Iraqis; that is, we must challenge our conventional understandings of response and focus on the ethical dimension, engaging with other people as fellow human beings. This model of response does not aim toward a definitive resolution: it does not orient itself toward a situation in which justice can finally and decisively be proclaimed in Iraq (and by whom exactly?). Instead, ethics as response advocates a daily attentiveness toward other people in all the delicate and minor ways we have at our disposal. This attentiveness is even more important when we are faced with pressing ethical responsibilities in Ukraine and other nations in the midst of bloody war. The immediacy of the horror when we witness the large-scale annihilation of life drives us to focus our attention upon the bloodiest conflicts; yet this simultaneously allows us to disengage from the lingering implications endured by those living within the former conflict zones. The violence experienced by Iraqis on a daily basis may not be as ferocious or pervasive as that in Ukraine, but it, nonetheless, demands attention and calls for a response.

Ethics as response, in this respect, proposes that non-Iraqis remain invested in the lives of Iraqi people: their politics and society. It asks us to keep thinking about ways in which we can help Iraqis build a new understanding of their society in which violent separation is not viewed as the primary objectivity of politics. Such an approach relies upon sustained effort and humility. It starts from the acknowledgement that non-Iraqis do not know what is best for Iraq or Iraqis. But the ideal of ethics as response remains hopeful that we can perhaps help Iraqis, in some ways and means, to negotiate new relationships between each other, their communities and socio-political divisions. This is not an easy task because we are faced with a context that has been shaped by decades of war, fear and distrust; we are starting from a position in which Iraqis have felt a pervasive insecurity for a very long time; and building trust between Iraqis, let alone between Iraqis and the outsiders who have decimated their society during the last two decades, will be extremely difficult. But this is a worthy goal because it aims toward the reduction of human suffering. The task of formulating specific responses to the current problems faced by Iraqis is not within the scope of this book. In fact, it should not be, because ethics as response is rooted in the belief that responses must be negotiated with other people and, importantly, Iraqis themselves. It is not for any-'one', let alone me, to prescribe how people should respond to Iraqis. Rather, this book is a call for other people to re-engage with Iraqis and the broader questions posed by war in their own small and incremental ways. This work has attempted to provide discussions on Iraq with the possibility of a fresh start. It is up to collectives of other people to negotiate how we can use this fresh start to begin to think about how we can help Iraqis in a more practical way.

The goal of this project has been to offer an alternative way of thinking about the relationship between violence and justice. I do not want, or expect, a wholesale shift to happen fully or in a single swoop. I do not expect any set of people to immediately reject the conventional assumptions that have been deeply engrained in our understandings of war. Instead, I am hopeful that the questions I have raised about the ways we routinely talk about and justify violence will have some small ripples within discourses on war. This is the starting point to what I would like to be a long engagement with others who are interested in questions of war, and an opportunity to rethink some of the assumptions that we often ignore. Above all, the preceding argument asks us to remain troubled by the justification of violence; not in the sense of a definitive rejection, because this is also tantamount to an ethico-political disengagement, but as a nagging reminder that even the most noble and morally palatable deployments of violence risk adversely affecting others in ways we do not intend or foresee. Ethics as response is a call for people to remain troubled and remain engaged in what it means to deploy violence in the name of justice.

Bibliography

Allam, Hannah (2003) 'Bremer Links Terrorists, Iraq Attacks', *Lawrence Journal-World*, 3 August, http://news.google.com/newspapers?id=D5wyAAAAIBAJ&sjid=8-gFAAA AIBAJ&pg=2575,699354&dq=paul+bremer&hl=en

Allawi, Ali A. (2007) *The Occupation of Iraq: Winning the War, Losing the Peace* (Yale: Yale University Press)

Al-Rahim, Ahmed (2005) 'The Sistani Factor', *Journal of Democracy*, 16(3), pp.50–53

Anderson, Perry (1983) *In the Tracks of Historical Materialism* (London: Verso)

Arango, Tim (2013) 'Some Iraqis Doubt Benefits of First Vote Since U.S. Departure', *New York Times*, 20 April, http://www.nytimes.com/2013/04/21/world/middleeast/iraqs-first-vote-since-us-exit-is-mostly-calm.html?_r=0

Arendt, Hannah (1963) *On Revolution* (New York: Penguin Books)

Arendt, Hannah (1970) *On Violence* (New York: Harcourt, Brace, Jovanovich)

Arraf, Jane (2003) 'U.S. Dissolves Iraqi Army, Defense, and Information Ministries', *CNN*, 23 May, http://edition.cnn.com/2003/WORLD/meast/05/23/sprj.nitop.army.dissolve/

The Art of Theory (2012) 'Interview with Michael Walzer', http://www.artoftheory.com/michael-walzer-the-art-of-theory-interview/

Asad, Talal (2007) *On Suicide Bombing (Wellek Library Lectures S.)* (New York: Columbia University Press)

Ashley, Richard K. (1984) 'The Poverty of Neo-Realism', *International Organisation*, 38(2), pp.225–286

BBC News (2003) 'Saddam's Symbol Tumbles Down', 9 April, http://news.bbc.co.uk/1/hi/world/middle_east/2933105.stm

Bellamy, Alex (2006) *Just Wars: From Cicero to Iraq* (Cambridge: Polity Press)

Bellamy, Alex (2008) *Responsibility to Protect* (Cambridge: Polity Press)

Benjamin, Walter (1999) *Illuminations*, ed. Hannah Arendt, trans. Harry Zorn (London: Pimlico)

Benjamin, Walter (2004) *Selected Writings: Volume 1, 1913–1926*, ed. Marcus Bullock and Michael W. Jennings (Cambridge, MA: Harvard University Press)

Boyle, Joseph (1997) 'Just and Unjust Wars: Casuistry and the Boundaries of the Moral World', *Ethics and International Affairs*, 11(1), pp.83–98

Bremer, Paul (2003) 'PBS Newshour Interview', *PBS*, 24 September, http://www.pbs.org/newshour/bb/middle_east/july-dec03/bremer_9-24.html

Brough, Michael W., Lango, John W. and van der Linden, Harry (2012) *Rethinking the Just War Tradition* (Albany: SUNY University Press)

Bull, Hedley (1979) 'Recapturing the Just War for Political Theory', *World Politics*, 31(4), pp.588–599

Bull, Hedley (2002) *The Anarchical Society: A Study of Order in World Politics*, 3rd Edition (New York: Columbia University Press)

Bulley, Dan (2009) *Ethics as Foreign Policy: Britain, the EU and the Other* (London: Routledge)

Burnham, Gilbert *et al.* (2006) 'The Human Cost of War: A Mortality Study 2002–2006', http://web.mit.edu/cis/pdf/Human_Cost_of_War.pdf

Bush, George W. (2002) 'Address to the UN General Assembly on Iraq', 12 September, http://edition.cnn.com/2002/US/09/12/bush.transcript/

Bush, George W. (2003) 'Speech to the American Enterprise Institute', 26 February, http://teachingamericanhistory.org/library/index.asp?document=663

Butler, Judith (2010) *Frames of War: When is Life Grievable?* (London: Verso)

Campbell, David (1998a) *National Deconstruction: Violence, Identity and Justice in Bosnia* (Minneapolis: University of Minnesota Press)

Campbell, David (1998b) *Writing Security: United States Foreign Policy and the Politics of Identity* (Minneapolis: University of Minnesota Press)

Chan, Sewell (2004) 'U.S. Civilians Mutilated in Iraq Attack: 4 Die in Ambush; 5 Soldiers Killed by Roadside Blast', *Washington Post*, 1 April

Chandrasekaran, Rajiv (2008) *Imperial Life in the Emerald City: Inside Baghdad's Green Zone* (London: Bloomsbury Publishing)

Chartrand, Molinda M. and Siegal, B. (2007) 'At War in Iraq and Afghanistan: Children in US Military Families', *Ambulatory Pediatrics*, 7, pp.1–2

Chomsky, Noam (2004) *Hegemony or Survival: America's Quest for Global Dominance* (London: Penguin)

CNN (2003) 'Saddam Statue Toppled in Central Baghdad', 9 April, http://edition.cnn.com/2003/WORLD/meast/04/09/sprj.irq.statue/

Cockburn, Patrick (2007) *The Occupation: War and Resistance in Iraq* (London: Verso)

Cole, Juan (2003) 'The United States and Shi'ite Religious Factions in Post-Ba'athist Iraq', *The Middle East Journal*, 57(4), pp.543–566

Corey, David D. and Charles, J. Daryl (2011) *The Just War Tradition: An Introduction* (Wilmington, DE: ISI Books)

Dawisha, Adeed (2009) *Iraq: A Political History* (Princeton, NJ: Princeton University Press)

Dawisha, Adeed and Diamond, Larry (2006) 'Iraq's Year of Voting Dangerously', *Journal of Democracy*, 17(2), pp.89–103

Demers, Anne (2009) 'The War at Home: The Consequences of Loving a Veteran of the Iraq and Afghanistan Wars', *The Internet Journal of Mental Health*, 6(1), pp.1–7

Derrida, Jacques (1978) *Writing and Difference*, trans. Alan Bass (London: Routledge)

Derrida, Jacques (1981a) *Dissemination*, trans. Barbra Johnson (London: The Athlone Press)

Derrida, Jacques (1981b) *Positions*, trans. Alan Bass (London: The Athlone Press)

Derrida, Jacques (1986) 'Declarations of Independence', *New Political Science*, 7(1), pp.7–15

Derrida, Jacques (1988) *Limited Inc*, trans. Samuel Webber (Evanston, IL: Northwestern University Press)

Derrida, Jacques (1992) *Acts of Literature*, ed. Derek Attridge (London: Routledge)

Derrida, Jacques (1997) *Of Grammatology*, trans. Gayatri Chakravorty Spivak, Corrected Edition (Baltimore: John Hopkins University Press)

Derrida, Jacques (2000) *Of Hospitality*, trans. Rachel Bowlby (Stanford, CA: Stanford University Press).

Derrida, Jacques (2002a) *Acts of Religion*, ed. Gil Anidjar (London: Routledge)

Derrida, Jacques (2002b) *Negotiations: Interventions and Interviews 1971–2001*, ed. and trans. Elizabeth Rottenberg (Stanford, CA: Stanford University Press)

Derrida, Jacques (2008) *The Gift of Death and Literature in Secret*, trans. David Wills (Chicago: The University of Chicago Press)

Derrida, Jacques (2009) *On Cosmopolitanism and Forgiveness*, trans. Mark Dooley and Michael Hughes (New York: Routledge)

Descartes, René (1998) *Meditations and Other Metaphysical Writings*, trans. Desmond M. Clarke (London: Penguin Books)

Dick, Kirby and Kofman, Amy Ziering, dir. (2003) *Derrida* (Jane Doe Films)

Elshtain, Jean Bethke (1992) 'Introduction' in Jean Bethke Elshtain, ed. *Just War Theory* (Oxford: Blackwell)

Elshtain, Jean Bethke (2002) 'A Just War?' *Boston Globe*, 6 October, http://www.boston.com/news/packages/iraq/globe_stories/100602_justwar.htm

Elshtain, Jean Bethke (2003) *Just War Against Terror: The Burden of American Power in a Violent World* (New York: Basic Books)

Elshtain, Jean Betkhe *et al.* (2002) 'What We're Fighting For: A Letter From America', *Institute for American Values*, http://www.americanvalues.org/html/wwff.html

Fick, Nathaniel (2007) *One Bullet Away: The Making of a Marine Officer*, Paperback Edition (London: Orion Books)

Filkins, Dexter (2009) *The Forever War* (New York: Vintage)

Finkel, David (2011) *The Good Soldiers* (London: Atlantic Books)

Foot, Christopher *et al.* (2004) 'Economic Policy and Prospects in Iraq', *Journal of Economic Perspectives*, 18(3), pp.47–70

Gaiman, Neil (2006) *Fragile Things* (London: Headline Publishing Group)

Gaiman, Neil (2010) *The Sandman Volume 2: The Doll's House* (New York: DC Comics)

Gerecht, Reuel Marc (2001) 'Liberate Iraq', *The Weekly Standard*, 14 May

Goldwin, Matthew (2012) 'Political Inclusion in Unstable Contexts: Mqutada al-Sadr and Iraq's Sadrist Movement', *Contemporary Arab Affairs*, 5(3), pp.448–456

Gregory, Derek (2004) *The Colonial Present: Afghanistan, Palestine, Iraq* (Oxford: Blackwell)

Grotius, Hugo (2010) *The Rights of War and Peace: Including the Law of Nature and of Nations (1901)*, trans. Archibald Colin Campbell (Whitefish, MO: Kessinger Publishing)

Gutmann, Matthew and Lutz, Catherine Anne (2010) *Breaking Ranks: Iraq War Veterans Speak Out Against the War* (Berkeley: University of California Press)

Harvey, David (2005) *The New Imperialism (Clarendon Lectures in Geography and Environmental Studies)* (Oxford: Oxford University Press)

Hashim, Ahmed (2003) 'The Insurgency in Iraq', *Small Wars and Insurgencies*, 14(3), pp.1–22

Heidegger, Martin (1996) *Being and Time*, trans. Joan Stambaugh (New York: State University of New York Press)

Heidegger, Martin (1998) *Pathmarks (Texts in German Philosophy)*, ed. William McNeil (Cambridge: Cambridge University Press)

Hendrickson, David C. (1997) 'In Defence of Realism: A Commentary on Just and Unjust Wars', *International Affairs*, 11(1), pp.19–54

Hennessey, Patrick (2010) *The Junior Officers' Reading Club* (London: Penguin Books)

Herring, Eric and Rangwala, Glen (2006) *Iraq in Fragments: The Occupation and its Legacy* (Ithaca: Cornell University Press)

Hersh, Seymour (2003) 'Selective Intelligence: Donald Rumsfeld has his own Special Sources. Are they Reliable?', *The New Yorker*, 12 May

Hiro, Dilip (1991) *The Longest War: The Iran–Iraq Conflict* (New York: Routledge)

Hirst, Aggie (2013) *Leo Strauss and the Invasion of Iraq: Encountering the Abyss* (Abingdon, Oxon: Routledge)

Hodge, Charles W. *et al.* (2004) 'Combatant Duty in Iraq and Afghanistan. Mental Health Problems, and Barriers to Care', *The New England Journal of Medicine*, 351(1), pp.13–22

Hodge, Joanna (2001) *Heidegger and Ethics* (London: Routledge)

Holmes, Jonathan (2007) *Fallujah: Eyewitness Testimony from Iraq's Besieged City* (London: Constable)

Holzgrefe, J. L. and O' Keohane, Robert (2003) *Humanitarian Intervention: Ethical, Legal and Political Dilemmas* (Cambridge: Cambridge University Press)

Honig, Bonnie (1991) 'Declarations of Independence: Arendt and Derrida on the Problem of Founding a Republic', *The American Political Science Review*, 85(1), pp.97–113

Hoyt, Mike and Palatella, John (2007) *Reporting Iraq: An Oral History of the War by the Journalists who Covered It* (New York: Columbia Journalism Review)

Jamail, Dahr (2008) *Beyond the Green Zone: Dispatches from an Unembedded Journalist in Occupied Iraq* (Chicago: Haymarket)

Johnson, James Turner (2001) *Morality and Contemporary Warfare* (Yale: Yale University Press)

Johnson, James Turner (2011) *Ethics and the Use of Force: Just War in Historical Perspective* (Surrey: Ashgate Press)

Johnson, James Turner (2013) 'Contemporary Just War Thinking: Which Is Worse, to Have Friends or Critics?' *Ethics and International Affairs*, 27(1), pp.25–45

Johnson, James Turner (2014) *Sovereignty: Moral and Historical Perspectives* (Washington, DC: Georgetown University Press)

Kaldor, Mary (1999) *New and Old Wars: Organised Violence in the Global Era* (Cambridge: Polity Press)

Kaplan, Robert (1993) *Balkan Ghosts: A Journey Through History* (New York: Vintage)

Karon, Tony (2003) 'Behind the UN vote on Iraq', *Time Magazine*, 23 May, http://www.time.com/time/world/article/0,8599,454203,00.html

Katzman, Kenneth (2008) 'Iran's Influence in Iraq', in Steven Costel ed. *Surging out of Iraq* (New York: Nova Science Publishers)

Keegan, John (2010) *The Iraq War: The Military Offensive, from Victory in 21 Days to the Insurgent Aftermath* (London: Hutchinson)

Kelly, Matt (2003) 'Rumsfeld Plays Down Iraqi Resistance Despite Deadly Attacks', *KSDK News*, 19 June, http://www.ksdk.com/news/story.aspx?storyid=42595

Kierkegaard, Søren (2006) *Fear and Trembling*, ed. C. Stephen Evans and Sylvia Walsh, trans. Sylvia Walsh (Cambridge: Cambridge University Press)

Koontz, Theodore, J. (1997) 'Noncombatant Immunity in Michael Walzer's Just and Unjust Wars', *Ethics and International Affairs*, 11(1), pp.55–82

Lasseter, Tom and Allam Hannah (2004) 'Fiery sky over Fallujah as Marines push into Embattled City', *Seattle Times*, 9 November, http://seattletimes.com/html/nation-world/2002085790_iraq09.html

Levinas, Emmanuel (1996) 'Martin Heidegger and Ontology', trans. The Committee of Public Safety, *Diacritics*, 26(1), pp.11–32

Levinas, Emmanuel (1999) *Otherwise than Being: Or Beyond Essence*, trans. Alphonso Lingis (Pittsburgh: Duquesne University Press)

Levinas, Emmanuel (2008) *Totality and Infinity: An Essay on Exteriority*, trans. Alphonso Lingis (Pittsburgh: Duquesne University Press)

Lévi-Strauss, Claude (1974) *Structural Anthropology* (New York: Basic Books)

Linklater, Andrew (1982) *Men and Citizens in the Theory of International Relations* (London: Palgrave Macmillan)

Litz, Brett T. *et al.* (2009) 'Moral Injury and Moral Repair in War Veterans: A Preliminary Model and Intervention Strategy', *Clinical Psychology Review*, 29, pp.695–706

Malkasian, Carter (2006) 'The Role and Perceptions of Counter Insurgency: The Case of Western Iraq, 2004–2005', *Small Wars and Insurgencies*, 17(3), pp.367–394

Maude, Lieutenant General Sir Stanley (1917) 'The Proclamation of Baghdad', http://wwi.lib.byu.edu/index.php/The_Proclamation_of_Baghdad

McDowall, David (2003) *A Modern History of the Kurds* (London: I. B. Tauris)

McMahan, Jeff (2009) *Killing in War* (Oxford: Oxford University Press)

Mearsheimer, John and Walt, Stephen (2008) *The Israel Lobby and US Foreign Policy* (London: Penguin)

Mejia, Camilo (2008) *Road from Ar Ramadi: The Private Rebellion of Staff Sergeant Camilo Mejia – An Iraq War Memoir* (Chicago: Haymarket Books)

Mills, Sgt. Dan (2008) *Sniper One* (London: Penguin)

Nancy, Jean-Luc (1991) *The Inoperative Community*, ed. Peter Connor (Minneapolis: University of Minnesota Press)

Napoleoni, Loretta (2005) *Insurgent Iraq: Al Zarqawi and the New Generation* (London: Constable)

Nardin, Terry (1997) 'Just and Unjust Wars Revisited', *Ethics and International Affairs*, 11(1), pp.19–53

Nasr, Vali (2004) 'Regional Implications of the Shi'a Revival in Iraq', *Washington Quarterly*, 27(3), pp.5–24

Nazir, Muntazra (2006) 'Democracy, Islam and Insurgency in Iraq', *Pakistan Horizon*, 59(3), pp.47–65

New York Times (2003) 'U.S. Troops Topple Hussein Statue in Central Baghdad', 9 April, http://www.nytimes.com/2003/04/09/international/worldspecial/09WIRE-CENTER.html

O' Driscoll, Cian (2008) *Renegotiation of the Just War Tradition and the Right to War in the Twenty-First Century* (London: Palgrave Macmillan)

O'Driscoll, Cian, Lang, Anthony Jr. and Williams, John (2013) *Just War: Authority, Tradition, and Practice* (Washington, DC: Georgetown University Press)

Olsen, Florian (2011) ' *"Those About to Die Salute You"*: Sacrifice, the War in Iraq and the Crisis of the American Imperial Society', *Geopolitics*, 16, pp.410–437

Orend, Brian (2000) *Michael Walzer on War and Justice* (Cardiff: University of Wales Press)

Patočka, Jan (1996) *Heretical Essays in the Philosophy of History*, ed. James Dodd, trans. Erazim Kohák (Chicago: Open Court)

Pattison, James (2012) *Humanitarian Intervention and the Responsibility to Protect: Who Should Intervene?* (Oxford: Oxford University Press)

Petraeus, General David H. *et al.* (2007) *The U.S. Army and Marine Corps Counterinsurgency Field Manual* (London: Chicago University Press)

Pfiffner, James (2010) 'US Blunders in Iraq: De-Baathification and Disbanding the Army', *Intelligence and National Security*, 25(1), pp.76–85

Pilkington, Ed (2013) 'US Military Struggling to Stop Suicide Epidemic among War Veterans', *The Guardian*, 1 February, http://www.theguardian.com/world/2013/feb/01/us-military-suicide-epidemic-veteran

Pin-Fat, Véronique (2005) 'The Metaphysics of the National Interest and the "Mysticism" of the Nation State: Reading Hans J Morgenthau', *Review of International Studies*, 31(2), pp. 217–236

Pin-Fat, Véronique (2010) *Universality, Ethics and International Relations: A Grammatical Reading* (London: Routledge)

Plato (1987) *The Republic*, trans. Desmond Lee (London: Penguin Books)

Politi, Daniel (2013) 'Obama: Syria "Would not be another Iraq or Afghanistan"', *Slate*, 7 September, http://www.slate.com/blogs/the_slatest/2013/09/07/barack_obama_weekly_address_president_makes_case_for_syria_strike_ahead.html

Powell, Colin (2003) 'Speech to the UN on Iraq', 5 February, http://www.washingtonpost.com/wp-srv/nation/transcripts/powelltext_020503.html

Project for the New American Century (1998), 'Letter to Bill Clinton on Iraq', http://www.informationclearinghouse.info/article5527.htm

Quigley, Carroll (1979) *The Evolution of Civilisations: An Introduction to Historical Analysis*, 2nd Edition (Indianapolis: Liberty Fund, Inc.)

Rahimi, Babak (2004) 'Ayatollah Ali Al-Sistani and the Democratization of Post-Saddam Iraq', *Middle East Review of International Affairs*, 8(4), pp.12–19

Rawls, John (1999) *A Theory of Justice*, Revised Edition (Cambridge, MA: Harvard University Press)

Ricks, Thomas, E. (2007) *Fiasco: The American Military Adventure in Iraq* (London: Penguin)

Ricoeur, Paul (2007) *History and Truth*, trans. Charles A. Kelbley (Evanston, IL: Northwestern University Press)

Robinson, Glenn (2007) 'The Battle for Iraq: Islamic Insurgencies in Comparative Perspective', *Third World Quarterly*, 28(2), pp.261–273

Rodin, David (2005) *War and Self-Defense* (Oxford: Oxford University Press)

Rubin, Barry (2005) 'Reality Bites: The Impending Logic of Withdrawal from Iraq', *Washington Quarterly*, 28(2), pp.67–80

Schmidt, Brian and Williams, Michael C. (2008) 'The Bush Doctrine and the Iraq War: Neoconservatives Versus Realists', *Security Studies*, 17(2), pp.191–220

Shadid, Anthony (2006) *Night Draws Near: Iraq's People in the Shadow of America's War* (London: Picador)

Shanahan, Rodger (2004) 'Shia Political Development in Iraq: The Case of the Islamic Dawa Party', *Third World Quarterly*, 25(5), pp.943–954

Shy, John (1976) *A People Numerous and Armed: Reflections on the Military Struggle for American Independence* (New York: Oxford University Press)

Sirkeci, Ibrahim (2005) 'War in Iraq: Environment of Insecurity and International Migration', *International Migration*, 43(4), pp.197–214

Smith, Michael Joseph (1997) 'Growing up with Just and Unjust Wars: An Appreciation', *International Affairs*, 11(1), pp.3–18

Smith, Steve (2004) 'Singing Our World into Existence: International Relations Theory and September 11', *International Studies Quarterly*, 48(3), pp.499–515

The United States Declaration of Independence (1776) www.constitution.org/usdeclar.pdf

US Department of Defense (2002) 'DoD News Briefing – Secretary Rumsfeld and Gen. Myers', 12 February, http://www.defense.gov/transcripts/transcript.aspx?transcriptid=2636

Vaughan-Williams, Nick (2005) 'International Relations and the Problem of History', *Millennium Journal of International Studies*, 34(1), pp.115–136

Wallace, David Foster (2005) 'This is Water: Some Thoughts Delivered on a Signification Occasion, About Living a Compassionate Life', *Keynote Speech to 2005 Graduating Class at Kenyon College*, http://grahamteach.com/wp-content/uploads/2011/07/This-Is-Water.pdf

Walzer, Michael (1973) 'Political Action: The Problem of Dirty Hands', *Philosophy and Public Affairs*, 2(2), pp.160–180

Walzer, Michael (1980) 'The Moral Standing of States: A Response to Four Critics', *Philosophy and Public Affairs*, 19(3), pp.209–229

Walzer, Michael (1983) *Spheres of Justice: A Defence of Pluralism and Equality* (Oxford: Blackwell)

Walzer, Michael (1987a) *Interpretation and Social Criticism* (Cambridge, MA: Harvard University Press)

Walzer, Michael (1987b) 'Notes on Self-Criticism', *Social Research*, 54(1), pp.33–43

Walzer, Michael (1990) 'The Communitarian Critique of Liberalism', *Political* Theory, 18(1), pp.6–23

Walzer, Michael (1994) *Thick and Thin: Moral Argument at Home and Abroad* (Notre Dame, IN: Notre Dame University Press)

Walzer, Michael (1996) *What It Means to be American: Essays on the American Experience* (New York: Marsilio)

Walzer, Michael (2005) *Arguing About War*, Paperback Edition (Yale: Yale University Press)

Walzer, Michael (2006a) *Just and Unjust Wars: A Moral Argument with Historical Illustrations*, 4th Edition (New York: Basic Books)

Walzer, Michael (2006b) 'Regime Change and Just War', *Dissent*, 53(3), pp.103–108

Walzer, Michael (2008) 'On Promoting Democracy', *Ethics and International Affairs*, 22(4), pp.351–355

Walzer, Michael (2012) 'The Aftermath of War: Reflections on *Jus Post Bellum*', in Eric Patterson ed. *Ethics Beyond War* (Washington, DC: Georgetown University Press)

Weiss, Thomas (2012) *Humanitarian Intervention (War and Conflict in the Modern World)* (Cambridge: Polity Press)

Wheatley, Steven (2006) 'The Security Council, Democratic Legitimacy, and Regime Change in Iraq', *The European Journal of International Law*, 17(3), pp.531–551

Wheeler, Nicholas J. (2000) *Saving Strangers: Humanitarian Intervention in International Society* (Oxford: Oxford University Press)

White, Hayden (1987) *The Content of the Form: Narrative Discourse and Historical Representation* (Baltimore and London: Johns Hopkins University Press)

Willson, Richard Ashby (2005) *Human Rights in the 'War on Terror'* (Cambridge: Cambridge University Press).

Worldpublicopinion.org (2006) 'Poll of Iraqis: Public Wants Timetable for US Withdrawal, but Thinks US Plans Permanent Bases in Iraq', 31 January, http://www.worldpublico pinion.org/pipa/articles/brmiddleeastnafricara/165.php

Wright, Evan (2005) *Generation Kill: Living Dangerously on the Road to Baghdad with the Ultraviolent Marines of Bravo Company*, Corgi Edition (London: Transworld Publishers)

Youseff, Maisaa (2008) 'Suffering Men of Empire: Human Security and the War on Iraq', *Cultural Dynamics*, 20(2), pp.149–166

Zehfuss, Maja (2007) *Wounds of Memory: The Politics of War in Germany* (Cambridge: Cambridge University Press)

Zehfuss, Maja (2011) 'Targeting: Precision and the Production of Ethics', *European Journal of International Relations*, 17(3), pp.543–556

Index